LITERACY AND NUMERACY IN LATIN AMERICA

Latin American Literacy and Numeracy Studies (LAL^{N}S) are fairly unknown in other parts of the world. This book charts new directions and explores the relationship between LAL^{N}S and international perspectives. Calling upon social practice approaches, New Literacy Studies, sociolinguistics, discourse analysis, and other paradigms, the contributors identify both convergent and divergent literacy and numeracy issues within the region as well as beyond the Latin American context.

Literacy and Numeracy in Latin America moves the field forward by bringing LAL^{N}S into wider focus and helping readers to understand the synergy with work from other perspectives and from other parts of the world and the implications for theory and practice. A lack of translated work until now between Latin America and, in particular, the UK, US, and Europe, has meant that such important overlaps between areas of study have gone unappreciated. In this way this volume is the first of its kind, a significant and original contribution to the field.

Judy Kalman is Professor, Departamento de Investigaciones Educativas, Centro de Investigación y Estudios Avanzados del IPN, Mexico City.

Brian Street is Emeritus Professor of Language in Education at King's College, London.

LITERACY AND NUMERACY IN LATIN AMERICA

Local Perspectives and Beyond

Edited by Judy Kalman
Brian Street

Routledge
Taylor & Francis Group
NEW YORK AND LONDON

First published 2013
by Routledge
711 Third Avenue, New York, NY 10017

Simultaneously published in the U.K.
by Routledge
2 Park Square, Milton Park, Abingdon, Oxon OX14 4RN

Routledge is an imprint of the Taylor & Francis Group, an informa business

Library of Congress Cataloging in Publication Data
Literacy and numeracy in Latin America: local perspectives and beyond / edited by Judy Kalman, Brian V. Street.
p. cm.

Includes bibliographical references and index.
1. Literacy—Latin America. 2. Numeracy—Latin America.
3. Education—Social aspects—
Latin America. I. Kalman, Judy. II. Street, Brian V.
LC155.A2L58 2012
379.2'4098—dc23 2011045405

ISBN: 978-0-415-89609-2 (hbk)
ISBN: 978-0-415-89610-8 (pbk)
ISBN: 978-0-203-80779-8 (ebk)

Typeset in Bembo
by Cenveo Publisher Services

Brian would like to dedicate this volume to his growing family—three children: Chloe and her husband Jon; Alice and her husband Jamie along with their young son Robinson; and Nicholas.

Judy dedicates this book to her husband Odón and their kids: Benjamin and Jessica, Rebeca and David.

CONTENTS

FOREWORD

Luis C. Moll

It is with pleasure that I write these lines as Foreword to this rich, varied, and important volume. This collection represents an unusual undertaking, as work done for different reasons in literacy and mathematics, and in the assorted circumstances that people inhabit throughout the Americas, is provided for the first time to English-language readers. This is bold work, challenging, as it does, deeply entrenched views of literacy and mathematics by offering a dynamic, practice-oriented, sociocultural perspective, grounded in a New Literacy Studies approach. The authors, however, also integrate a broad range of ideas into their work, ranging from the writings of Bourdieu and Foucalt to those of Vygotsky and Wittgenstein.

Doing research in Latin America from such a perspective is ultimately about, as Marinho (this volume) writes about Brazil, working with diversity, with differences, and with the many social contradictions that characterize life in the region. I found the research profoundly cultural–historical in its emphasis on the paramount importance of social context, and in studying real people, doing real things, in real settings for living. The focus, then, is not on models of culture that establish norms that differentiate entire groups of people, say, Brazilians from Mexicans and from Argentineans, that would have been an essentialist trap, I am sure. Instead, the focus is on how people "live culturally," to borrow a phrase from Ingold (1994), within their particular and varied circumstances of life, using the knowledge, social practices, and cultural resources at their disposal. With such an emphasis, issues of identity, and emotions abound, both in the plurality of practices studied under challenging circumstances, and in the formation of fundamentally new arrangements for learning with those who are dispossessed.

This focus on diversity also means that, methodologically, the objective is to understand cultural practices not by some sort of detached or impersonal observation, but by interacting with, by engaging, and documenting the multiple lived experiences of the people with whom one works and studies. The focus is also on questioning power and domination by revealing the ideologies of uniformity and control that still predominate in, and in many ways come to define, schooling practices in the region, and elsewhere for that matter. The reconfiguration of schooling can be deeply frustrating, as some chapters illustrate, it is a most robust and fluid social structure, as power adapts well to each new exigency, but opening its structure to life remains a critical task for development, especially for those challenging the social order, often at great risk to themselves.

Each study reported herein, therefore, depicts fluid events, a story in the making, and, even though these processes may be frozen on paper for examination, it is clear theoretically and methodologically that the ultimate goal is to understand culture in action, in movement, and, in this case, how literacy and numeracy are artifacts, things, for living and acting on the world. I close these brief remarks by borrowing again, this time at length, from Ingold (2006), who makes the point cogently:

> The conditions that enable scientists to *know*, at least according to official protocols, are such as to make it impossible for scientists to *be* in the very world of which they seek knowledge. Yet all science depends on observation, and all observation depends on participation—that is, on a close coupling, in perception and action, between the observer and those aspects of the world that are the focus of attention. If science is to be a coherent knowledge practice, it must rebuild on the foundation of openness rather than closure, engagement rather than detachment… Knowing must be reconnected with being, epistemology with ontology, thought with life.
>
> *(19, emphasis in original)*

The work of our colleagues in this volume exemplifies this openness and engagement, and reconnections to life.

References

Ingold, T. (1994). Introduction to culture. In T. Ingold (ed.), *Companion Encyclopedia of Anthropology: Humanity, Culture and Social Life*. London: Routledge, 329–49.

Ingold, T. (2006). Rethinking the animate, reanimating thought. *Ethnos* 71 (1), 9–20.

PREFACE

The authors and editors of this book describe literacy and numeracy in a variety of Latin American contexts. They attempt to link the specificity of these local studies with wider global issues, regarding the meanings and uses of literacy and of numeracy in contemporary societies and the policy and practice associated with them. Whilst such debates have been going on for some time in many areas of the world, the present volume is unique in locating them in the context of Latin America. The editors have in fact collaborated in the production of a book in Spanish also based in Latin America (J. Kalman, and B. Street (eds.) (2009), *Lectura, escrita y matematicas como practicas sociales: Dialogos con America Latina.* Siglo XXI editores and CREFAL) and some of the authors included in the present volume did also write for that book, but this is the first time that such a volume has been published in English. A major aim of such an enterprise is to make LAL^{N}S available to a much wider audience. It both establishes that there is indeed a field that can be called Latin American Literacy and Numeracy Studies, and also that work in this field links in important ways with work in other parts of the world and in particular reinforces the association of local and global that is now a central part of studies in literacy and numeracy.

The authors have studied a range of specific contexts in Latin America—notably Mexico, Brazil, Argentina—in order to understand the complexities of literacy and numeracy and have developed ideas and perspectives that can help us understand the deeper features of how reading, writing, and numbers are used and constructed in social interaction, institutional settings, schooling, and everyday life. Such a qualitative perspective allows us to dialogue and debate with local policy makers and officials as well as with high level operators in international agencies, drawing their attention to the distinctiveness of what is going on in

Latin American countries, whilst at the same time, locating these practices in the global context that such agencies have to focus upon as they attempt to improve standards across nations.

The editors' Introduction provides an overview of the themes addressed in the volume, which is organized in three sections. The title of each section indicates how a particular aspect of the field is qualified with reference to Latin American perspectives. In Part I, "Latin American Literacies: Theoretical and Methodological Approaches", three chapters address key aspects of the field: Chapter 2: "The Local and the Global in Literacy Practices in 'Traditional Communities'" by Marildes Marinho from Brazil; Chapter 3: "Access to Written Culture as Part of the Social Reproduction Strategies of Rural Families in Córdoba (Argentina)" by Elisa Cragnolino from Argentina; and Chapter 4: "A Multimodal Approach to the Understanding of Students' Collaborative Writing of Digital Texts" by Enna Carvajal from Mexico.

Part II includes four chapters that focus on the specifics of "Literacy and Numeracy as Social Practice: Latin American Perspectives": Chapter 5: "GPS Technology, Map Reading, and Everyday Location Practices in a Fishing Community" by Judy Kalman; Chapter 6: "When Illiterate isn't Illiterate. Reading Reality in a Multimodal Way" by María del Carmen Lorenzatti; Chapter 7: "Indexical Signs within Local and Global Contexts: Case Studies of Changes in Literacy Practices across Generations of Working Class Families in Brazil" by Maria Lucía Castanheira; and Chapter 8: "Survival of Original Knowledge" by Irma Rosa Fuenlabrada and María Fernanda Delprato.

Part III, "Literacy and Numeracy in Education: Experiences in Latin America," links the study of literacy and numeracy with work in education. This section comprises: Chapter 9: "When Literacy Brings Too Many Risks: A Successful Lesson in Failure by shirley Brice Heath and Daniel Sobol; Chapter 10: "The Brazilian Landless Movement and a Mathematics Education Research Program" by Gelsa Knijnik; Chapter 11: "Reading, Writing and Experience: Literacy Practices of Young Rural Students" by Gloria Hernández; Chapter 12: "Technology and Literacy: Towards a Situated Comprehension of a Mexican Teacher's Actions" by Irán Guerrero; Chapter 13: "Preambles, Questions, and Commentaries: Teaching Genres and the Oral Mediation of Literacy" by Elsie Rockwell; and Chapter 14: "Learning English in Mexico: Transnational Language Ideologies and Practices" by M. Sidury Christiansen and Marcia Farr.

Finally, these different studies and approaches are brought together in an Afterword by David Barton, from Lancaster University in England. As one of the leading figures in the international scene, he is able to provide an authoritative overview of exactly what these studies contribute, locating them both in their local Latin American context but also bringing out exactly how they can be seen from a wider international perspective. And that is the contribution of the volume, to bring these perspectives together and to provide the reader with rich detailed accounts of literacy and numeracy in the actual lives of people in different countries in that part of the world as well as drawing upon the wider global literature which in turn these studies are able to enhance and move on. Moving on is, indeed, the central point of this book and we hope that readers will be able to use it to enhance their own studies, and to move on in their own applied work in policy and practice.

ACKNOWLEDGMENTS

Thank you to Naomi Silverman at Routledge for taking such a positive interest in the project and supporting the production of this book.

We are indebted to Maria del Carmen Arriola for her hard work and painstaking revisions of each chapter; she also kept us organized and on top of corresponding with the many authors, helping to clarify doubts and ambiguities, secure needed information, and, in general, keeping the ball rolling. Also we would like to thank Alma Becerra at the Departamento de Investigación Educativa at Cinvestav in Mexico City for her follow-up on several details related to the book. Irán Guerrero came to our rescue on several occasions with technological advice and solutions; she also generously donated her time to participate in what turned out to be a meticulous revision of the bibliographical references. Tonatiuh Paz, an intern at the Laboratorio de Educación, Tecnología y Sociedad, also at the Departamento de Investigación Educativa, kept track of contributors' information. Maria Lucía Castanheira in Minas Gerais in Brazil was very helpful in locating some sources in Portuguese as well.

Thanks to all the authors for their hard work. We are very grateful to Luis Moll who wrote the Foreword for the book and to David Barton who wrote an Afterword—such additions by leading scholars help to place the volume in the larger context.

1

INTRODUCTION: LITERACY AND NUMERACY IN LATIN AMERICA

Local Perspectives and Beyond

Judy Kalman and Brian Street

Overview

In Latin America an idea that has dominated the official discourses for decades, if not more, is that reading and writing are singular, neutral, and objective skills that are learned through a progression of ordered exercises and then transferable to any situation. Consequently, governments and international agencies alike have promoted—and continue to promote—programs that reduce literacy to mechanical skills that fail to capture the complexity of literacy practices in the social world. But there is a growing research community of established and new scholars located in Latin American universities and research institutions who interact with colleagues from abroad and who together collectively question this view.

The authors and editors of this book have studied specific contexts in order to understand the complexities of literacy and numeracy and developed theoretical notions that help us understand the deeper features of how reading, writing, and numbers are used and constructed in social interaction, institutional settings, schooling, and everyday life. We have produced original studies that not only reveal the particularities of specific cases, but have also constructed a qualitative perspective that allows us to dialogue and debate with local policy makers and officials as well as high-level operators in international agencies.

For many of us working on this research agenda, the premises of New Literacy Studies (NLS) have provided important conceptual and methodological insights that have helped us to develop a discourse divergent from the representations of literacy and numeracy that we continuously encounter in local and international circles. NLS's vision of literacy and numeracy as social practices; its recognition of literacy and numeracy as multiple constructs (best thought of as literacies and numeracies); its acknowledgment of their immersion in ideologies, historical processes, and power relations that often reach beyond the immediacy of situated events; and, finally, the idea that literacy and numeracy are sustained by socially constructed epistemological principles (Barton and Hamilton, 1998; Collins and Blot, 2002; Kalman, 1999; Kalman and Street, 2009; Pahl and Rowsell, 2006; Street, 1984, 1993, 1995, 2005) have all been key in creating a broader, more complex, and nuanced version of how reading and writing are accomplished and how numeracy unfolds in particular contexts.

Despite the years of work in this field, Latin American Literacy and Numeracy Studies (LAL^{NS}) are fairly unknown in other parts of the world. For many, the first and perhaps only reference is Paulo Freire, whose important contributions to the field were mainly during the 1970s–1990s. The purpose of this book is to make the work of Latin American scholars available to academics and interested readers elsewhere, and to indicate new directions in the field of literacy and numeracy studies in Latin America, while exploring the relationship between these and international perspectives. The editors, as well as many of the authors writing in this volume, also published together in a previous volume in Spanish entitled *Lectura, escritura y matemáticas como prácticas sociales: Diálogos con América Latina*, published in Mexico City by Siglo XXI Editores and CREFAL. In this first collective work, we presented original research in Spanish and translated authors whose work is often unavailable to a Spanish- or Portuguese-reading public. In this current volume we seek to continue this first conversation and contribute to the understanding of literacy practices, the local specificities attributable to their Latin American contexts, as well as their place in broader global sites and institutions. By integrating NLS with other social practice perspectives, sociolinguistics, discourse analysis, social semiotics, and other paradigms, the papers presented in this volume identify both convergent and divergent literacy and numeracy issues within the region, as well as beyond the Latin American context.

While the chapters presented in this book are all situated in Latin America and share a common point of view regarding reading, writing, and numeracy, they cover a wide variety of situations—from the Landless Movement in Brazil to fishers on the Caribbean coast; from multimodal interpretation of writing in a senior high school in Mexico City and amongst adults in Argentina to the interpretation of official textbooks in public school classrooms in Brazil; from homework assignments to computer-mediated collaboration in a science class in Mexico. Together they provide a rich tapestry of literacies, of people reading and writing in a variety of ways, for a variety of purposes, and in a variety of complex social contexts. Similarly, we turn to other theories and research agendas to expand and enrich what NLS has offered us.

Another desired outcome of this work is to disseminate current theory and empirical research to a more extensive audience in contexts both of Latin America and beyond. Our intention is to move the field forward by bringing the work of LAL^{NS} into wider focus and to help readers to understand how the synergy with work from other perspectives and other parts of the world may both contribute to work on literacy and numeracy in Latin America, and also enable the Latin American perspective to contribute to the broader field of inquiry with its implications for theory and practice. We anticipate work in literacy and numeracy both in Latin America and elsewhere being enhanced by this encounter as researchers involved in the exploration of their relationship learn from each other. The lack of translated work until now between Latin America and, in particular, the U.K., U.S., and Europe has meant that such important overlaps between areas of study have gone unrecognized. In this sense this volume is the first of its kind and can justly claim to offer an original contribution to the field.

The chapters in this collection can be seen as making the following contributions:

- Presenting distinctive aspects of LAL^{NS} developed by researchers from Latin America and others working in Latin America, particularly those from the U.K. and U.S.
- Exploring the potential for social practice approaches to literacy and numeracy to contribute to LAL^{NS} with the aim that new directions will emerge for both theory and practice.
- Indicating the significance of work in Latin America for research and practice in other parts of the world.
- Framing research in Latin America in the context of theoretical developments and recent studies, both on local literacy and numeracy practices, and on applied programs in other parts of the world.
- Linking work in the field of literacy studies with recent work in other fields, specifically numeracy as a social practice and multimodality as a social semiotic resource.

Common Theories, Concepts, and Themes

The authors contributing to this volume share the NLS approaches in their analysis of reading, writing, and other meaning-making activities, through the study of events and practices that involve the use of texts. In this perspective, *all* uses of writing—on paper or on screen—"are seen as *social practices* in that they are constituted within and through social relations and are embedded in broader social contexts and processes" (see Rockwell, this volume). Collectively, they present a rich and complex portrait of literacies situated in their local contexts. Their analysis of the deeply social nature of reading and writing and its links to local and global realities, institutional and power relationships, situated practice, and schooling is enhanced by a variety of approaches, points of view, and theoretical positions, and their location in different countries of Latin America.

One issue addressed by most of the chapters is the multiple and often confounding terminology that surrounds literacy research in Latin America, a reflection of ongoing transformations in the field and longstanding conceptual debates.

The first chapter following this introduction is by Marildes Marinho, whose untimely death soon after completing her work for this volume has saddened us all (see comments the beginning of Chapter 2). Marinho notes the abundance of terms that surround literacy in Latin America—*alfabetização/alfabetización*, *letramento*, *letrisme*, written culture—and argues that they are not synonyms but different social (and often political) concepts of reading and writing, while others add the notion of *literacidad* (a neologism taken from the English word literacy) to the mix (see Lorenzatti's and Cragnolino's chapters in this volume). What underlies this discussion is a conceptual move that seeks to distance NLS in Latin America from the more restricted notion of *alfabetização/alfabetización* mentioned above; a brand of literacy that starts and stops with letters and sounds and is closely tied to official campaigns and international agencies. In its place is a more nuanced, analytically rich, and socially complex view of what people do when they read and write.

Another common thread in these studies is the recognition of the multiplicity of literacies; with the global outburst of multimodal media, several authors' research centers on how diverse forms of both printed and digital symbolic representations converge in interpreting and producing meanings. Following the work of Gunther Kress, Enna Carvajal, Irán Guerrero, Judy Kalman, and María del Carmen Lorenzatti, each contemplate different aspects and issues related to the appropriation of literacy, be it in or out of school. They note how intermodal interaction "is part of the production of meaning, which depends on the social contexts of use and the motivations and interests of the person(s) who produce(s) it" (Carvajal, this volume). Lorenzatti's contribution not only shows the importance of these motivations and interests, but underlines the social nature of privileged use. The telling title of her chapter, "When Illiterate isn't Illiterate", foreshadows a complex recounting of "how adults that did not go to school during childhood and never learned to read written language well, manage to build and give meaning to different semiotic representations occurring in their daily life." By accompanying Marta Graciela, a Bolivian immigrant residing in Argentina, through the day, she is able to observe her in a variety of literacy events and recognize how she uses oral language, writing, images, gestures, and objects in order to understand signs, wrappers, documents, and screens. Lorenzatti is able to reconstruct how adults like Marta Graciela interact with socially constructed signs without centering on writing but by adapting it to other modes, thus revealing the limited scope of a notion of literacy based on *alfabetización*. Marta Graciela is able to interpret a broad variety of texts by integrating written words into larger contexts, not by isolating them.

Given the extreme disparities and widespread poverty in the region, a third common theme—in addition to the issue of terminology and the significance of multimodal understandings—is the relationship of literacy to social class and economic status. Many of the authors do their research in contexts of poverty—the Landless Movement in Brazil, indigenous communities in Mexico, marginalized urban neighborhoods in a number of countries (See Knijnik, Castaneheira, Heath and Sobol, Marinho, Kalman, Lorenzatti, Farr, Fuenlabrada and Delprato, Cragnolino, this volume). Since the beginning of the 20th century—and still present today—part of the official discourse prevalent in the region has been the promise of literacy providing upward mobility. Two of the contributing authors who look closely at this issue use the work of Pierre Bourdieu, particularly the notions of social capital and habitus, to unearth the ways in which literacy does not necessarily contribute to social mobility, and may in fact be an obstacle for economic betterment. In their chapter, Heath and Sobol analyze an educational program in the Dominican Republic that sought to create conditions for young people to participate in the local labor market through a community vegetable cooperative, but its efforts were undone by local beliefs regarding desirable and undesirable work, making participation unattractive to local youths. In this case, even though the program situated multiple opportunities for literacy use in ongoing activities of planting, growing, caring for, and marketing produce, literacy failed to provide social capital to effect the participants' mobility. Likewise, Cragnolino argues that the notion of habitus "helps explain how and to what extent families in rural Argentina implement certain alternatives of action in relation to written culture and discard others, put certain mechanisms into effect and make efforts to employ them." Habitus, she notes, incorporates notions that lead them to think and to define "what is possible" and "what is impossible," "what is thinkable" and "what is unthinkable," "what is for us" and "what is not for us." In both cases, deep cultural beliefs about literacy, work, education, and the social world, shape what literacy is used for and what literacy may or may not help people to accomplish in their lives.

The Local and the Global: Literacy and Numeracy in a Connected (Social) World

Local contexts provide the scenarios for understanding how reading and writing are accomplished, taught and learned, disseminated and appropriated in ongoing everyday life. The repertoire of highly situated practices documented here spans using geographic positioning devices for lobster fishing, map reading, calculating costs of wares for a regional market, using digital technologies to set up a language school in a township, learning academic literacies, navigating bureaucracies, and tourism. In each case, the uses of literacy and numeracy are observed and the researchers try to understand why people do what they do (Geertz, 1983), and how what they believe and know shapes their actions (Barton, Hamilton, and Ivanic, 2000). In the context of international globalization, to say that local realities are somehow tied to global economic, political, and social tendencies is to state the obvious; however, the analyst's job is to uncover the specific mechanisms that connect the local to the global, be it through the analysis of Bourdieusean relations between habitus and field (see Grenfell, Bloome, Pahl, Rowsell, and Street for a forthcoming attempt to link NLS and Bourdieu), of policies and programs, shifts in local economies, shifts in demographics, or the incorporation of computers and connectivity into the classroom.

Changes in economic structure are illustrated by the shift from local exchange to national and international economic forces based on monetary systems. Issues around employment are a common denominator in Fuenlabrada and Delprato's discussion of the marketplace; in Castanheira's analysis of a working-class neighborhood in Brazil; and in Heath and Sobol's account of a community gardening and literacy development project in the Dominican Republic. Authors map the changes and their implications for literacy; in the case of Trombetas in Brazil, Castanheira maps changes from the 1980s to the present day in the kinds of employment available to unschooled and underschooled adults, changes which have important implications for household economies, and may determine if it is decided that children go to school or go to work. Shifts from the agricultural sector to the industrial and to the service sector may also bring changes in literacy requirements, which in principle would be supplied by schooling, although, as shown by several authors, this is not always the case. In Argentina, for instance, Cragnolino describes how older people recall their household duties and their children's roles in the family's economic wellbeing, while many of their children and grandchildren have moved to cities, to centralized work, and now need the associated literacies.

A recent trend emerging over the last few years is the return of immigrants from the United States to Mexico, and a slowing down in the flow of immigrants northward bound. This is due for the most part to the militarization of the Mexican–U.S. border, the recent restrictions in employment policy and the heavy sanctions levied against employers hiring undocumented workers in several border states, and programs for "voluntary return," i.e. leaving the U.S. without facing penalties for illegal entrance. However, this does not imply a complete severance between the U.S. and Mexican economies, as Farr and Christiansen show in their chapter. They document how the experience of immigration and what they refer to as transnationalism "is shaping parental language ideologies in Mexico that now emphasize the importance of acquiring both oral skills and literacy in English in non-traditional ways." Their research site is a town of 15,000 in the mountains of central Mexico with a high percentage of returnees from the U.S. They note that

> Those who migrated to the U.S. but then returned to this town introduced new possibilities that inevitably impacted traditional ideologies, even among those who have never been migrants themselves but who now can imagine themselves living and working in the U.S. In this town the presence of such people intensified a long-perceived link between English and more lucrative employment, changing local parents' traditional ideas and practices regarding 1) the importance of acquiring oral and written English, 2) processes of learning, and 3) bilingualism. Parents now place high priority on the acquisition of both oral skills and literacy in English.

In order to identify and examine the connections between the local and the global, several authors have used Kalman's (2001, 2004) notions of access and availability in order to signal the material aspects of literacy (the physical presence of print materials, infrastructure for reading and writing such as post offices, libraries, bookstores, news-stands, cybercafés, etc.) and their relation to the social conditions necessary for literacy learning. Within this distinction, availability refers to the provision of resources, such as computers, notebooks, etc., whilst access refers to the ways in which people actually take hold of such resources; through, for instance, interaction with other readers and writers, appropriation of discourses, the understanding of meanings, the construction of relevant forms of participation within a given context, as well as learning the intricacies of written conventions with reference to the contexts, purposes, and actors involved in literacy learning (Blommaert, 2005). So defined, access is useful for moving beyond the issue of "availability" and discussing the social and epistemological conditions necessary for literacy learning and use (see Hernández, Kalman, Guerrero, and also Castanheira, this volume). Access, then, refers to the social interactions, collaborations, and support for learning discourse practices. Marinho (this volume) poignantly illustrates how sitting in university lecture halls is not enough for indigenous students to learn and use formal academic discourse—beyond the availability of such discourse, the issue of how students actually gain access to it requires sharing, interacting, and negotiating with others seasoned and experienced in the demands of widespread and dominant academic practices. In schools in Mexico, and across Latin America, computers have been made available to students and teachers through distribution policies. Yet as both Carvajal's and Guerrero's chapters show, the availability of equipment and installed infrastructure (electricity, devices that work, connectivity, software) necessary for incorporating computers in classrooms is not enough—access is achieved only through interaction with more expert others who demonstrate how technologies work, share practitioners' knowledge, and collaborate with others through authentic purpose and use.

Castanheira also looks at the connections between global forces and local realities through the lenses of access and availability, showing how the lives and surroundings of immigrants from rural to urban areas have changed over the last twenty years. She presents data from ethnographic research on writing involving working-class families that live in Trombetas on the outskirts of Belo Horizonte, Brazil. This study contrasts intergenerational data from two case studies of writing in these families, calling upon the concept of indexical signs as a way of analyzing such intergenerational changes in both local and global terms. The first case study was developed in the late 1980s and the second is in development. In the last two years Castanheira has had the opportunity to interview members from two families that participated in the previous case study. Contrasting data from these two studies, she identifies changes that have taken place over the years and examines the implications of these changes for the meanings of literacy for the people living there. She finds that "the wider context affected changes in the local context and what goes on in the life of Trombetas' residents in relation to writing and reading." There are clear signs that index changes both in the availability of literacy and in the access to reading and writing took place locally in Trombetas. By looking for traces of change in public spaces, for instance in the increase in written signs in these spaces, the arrival of several neo-Pentecostal churches, and the distribution of religious texts, she is able to document local versions of larger and broader trends occurring across the country and in Latin America.

Literacy and Numeracy as Mediated Practices

Current thinking in scholarly research underlines the connected nature of literacy and numeracy to other forms of representation, artifacts, and cultural beliefs and practices. Each of these come into play in situations where reading and writing are mediated by different participants in situated events. Readers and writers and practical doers of mathematics serve as intermediaries for one and another and use different (physical or symbolic) tools and devices to mediate interactions, whether they be pencil and paper, computers or pottery.

Through the analysis of mediated events, issues of identity and epistemology are unveiled. Lorenzatti's research subject notes the importance of others in her uses of literacy in her life, noting "there is always someone who knows." The study looks at issues of mediation by using concepts from interactive sociolinguistics as well as multimodal theory to examine how reading is accomplished through articulating written text with gesture, talk, and other forms of representation. Lorenzatti notes that in official discourse (and in popular belief), "underschooled adults and youths are believed to be incapable of having an opinion, understanding complex issues and participating in social spaces." Through her analysis of one woman's interpretive practices, she presents a counter-example of how one woman interprets complex multimodal representations and participates actively and competently in a variety of social contexts. Lorenzatti insists that "People tagged and classified as illiterate, uneducated, or poorly educated are generally not recognized as active social subjects despite the fact that they participate in institutions, struggle for their rights, constitute families or plan and develop projects for the future."

The theme of how people get by and use literacy and numeracy, often through alternative practices, recurs in the writing of other authors who work in marginalized communities. At the center of Irma Fuenlabrada and María Fernanda Delprato's chapter is the issue of cultural mediation of numeracy practices and the predominance of community values over market notions of maximum profit. Through their exploration of issues of identity and epistemology, they clearly illustrate how mathematics is a situated cultural practice. They analyze how an indigenous woman's traditional beliefs and practices work as cultural mediators in her uses of numeracy. Elisa is an indigenous P'urhépecha from Michoacán

(Mexico) with little formal schooling and the leader of a self-reliant organization that hand-produces textiles and pottery in her community. Her leadership role within her community positions her as an intermediary between "outsiders" and "insiders" of her organization, and from this perspective she understands the outside knowledge that regulates trade exchanges and learns to survive them. She assumes the value of traditional production, as well as communal organizational practices, and defends the use of her local language while learning enough Spanish to negotiate competently with potential buyers and intermediaries. She is respectful of local cultural practices regarding shared wellbeing, fairness, and equality of opportunities. Consequently, in order to understand the mathematical logic behind the solution strategies of the arithmetical situations that she confronts, it is necessary to grasp the social concerns that support them and construct the link between the appropriation of specific numeracy practices and the identity processes that mediate Elisa's learning.

Local culture and the interconnections of race, class, and occupation also mediate how local communities participate in organized educational endeavors. Heath and Sobol show, with reference to research in the Dominican Republic, how programs based on the best learning theories available may fail because planners ignore important cultural dimensions of learners' willingness (or lack of it) to participate. Neither intellectual resources, nor carefully designed programs, nor the best possible match between outside personnel and local learners can overcome the reliance of learners on the security they find in the traditional values and assumptions of their community and the everyday interactions of their current social status. Social mobility depends, among other things, on changes in both education and employment patterns, but individuals face extraordinary risk in either acquiring literate habits or foreseeing a future different from that of family members. The chapter examines a case of "failure" within a framework detailing the interplay of local social structures with external charity organizations, local benefactors, and enlightened school developers.

What locals choose to learn and decide to use of their newly appropriated knowhow is also mediated by cultural practices and local knowledge. Kalman's chapter continues this theme through an examination of fishers' encounters with marine scientists and describes how they worked together to chart underwater benthic resources. Through this contact, the fishers living on the coast of Quintana Roo (Mexico) were able to watch the scientists use hand-held Geographic Positioning Systems (GPS) devices and took an immediate interest in them, because they saw GPS as potentially useful for improving their fishing techniques. Riding together in the boats, the fishers and the scientists collaborated in an informal situation for learning with a particular epistemological slant. Kalman notes that

> through in situ demonstrations the researchers involved the fishers in experimental situations, creating on-the-spot opportunities to hypothesize

> about how the GPS works and the immediate conditions to validate or reject their conjectures … the marine scientists asked, "What do you think will happen to this line (on the screen) if you go over there?" They could then redirect the boat with the GPS in hand to see if their expectations were correct or not.

She also notes that, although the fishers were open to the scientists' ways of navigating, they refused to "shed" the knowledge constructed from years of working without technology and held on to practices deeply linked to their local identities. Knowledge from years of fishing and familiarity of local procedures were templates for understanding the hand-held GPS brought by scientists and "for incorporating them into their practice repertoires." Kalman concludes that "Fishers used their knowledge to decide when and how to use new practices and they determined in specific situations which one was more appropriate, when both were required, or traditional knowledge and know-how was more reliable."

Schooling; Teaching, Learning, and Experience

While schooling is often referred to as a homogeneous context, educational research in literacy and other fields has shown that what goes on at school is far from monolithic. Street (2005, 5) notes that "as in the case of other literacy practices, academic practices are socially constructed, vary across time and space and have to be learned in specific ways, just as the other literacies that people encounter across a life time." In the present book, several authors look at specific aspects of schooling in order to understand how literacy is constructed and how teachers' and students' beliefs and knowhow contribute to learning.

While the connections between orality and literacy are developed from several angles outside of school in both Lorenzatti's and Castanheira's chapters, Rockwell (this volume) takes this discussion into school by selecting specific moments in which teachers engage in verbal interaction with students while texts are taken up in the oral performance of reading. She notes that "during such sequences the distinction between oral and written modes becomes blurred". She goes on to observe:

> Within this social context, as most others, an oral/literate dichotomy does not hold. As elsewhere, uses of writing are context-bound, and their significance is a consequence of local culture and social relations.
>
> *(Rockwell, 2010).*

Her observations remind us that, as elsewhere, uses of writing are context-bound, and their significance is a consequence of local culture and social relations, interpretative practices and local knowledge. She notes how school literacy practices are embedded in oral discourse and that teachers model, guide, and evaluate students' relationship to the textbook lessons, and that that there are important differences from teacher to teacher in the way they verbally mediate the reading of written texts.

The themes of experience and epistemology are foregrounded in Gloria Hernández' study on the various ways in which young rural students' literacy experiences and practices in everyday contexts in a community in central Mexico influence their reading and writing practices at school. She centers her discussion on the contexts of knowledge building and argues that teachers' dialogues with students about the ways they learn about literacy and use reading and writing reduce their reading experiences to a static product, missing processes with multiple interactions and wrapped in larger complexities. She points out:

> What is common in the classrooms I have studied as well as others reported in the literature is the triggering question: what do you know about this? But not "how do you know it?" or "What do you think about it?" or "Where did you learn it?" or "With whom did you learn it?" With this, only two things are strengthened: on the one hand, repetition without reflection about the constructed knowledge, and on the other, the notion of experience only as a product, and not as a process with multiple interactions and wrapped in larger complexities.

As a result, Hernández found that, within her research site, the contexts for reading at school and constructing reading experience and the context of reading at home take separate paths due to the intrinsic meaning that these activities have in each specific space and the difficulties teachers and students have to somehow relate them.

This issue of how students and teachers know what they know is also salient in classrooms where digital technologies are being incorporated into teaching and learning. Although many believe that adding computers to classrooms results in immediate educational change, there is a growing body of literature that says otherwise (Kress, 2002; Kress and van Leeuwen, 2001; Lankshear and Knobel, 2006). As in many parts of Latin America, in Mexico the Secretary of Public Education has been distributing computers and modems throughout the school system and calling on the use of technology as a strategy for educational change and renovation. Enna Carvajal accompanied a group of three high school students using technology in a science project, and analyzed the ways in which they construct meaning when writing academic texts. As in the case of the other chapters presented here, contextual factors were key to fostering the collaborative construction of their written project. In ways similar to the use of multimodal resources by Marta Graciela, the protagonist of Lorenzatti's chapter (see above), for the students in Carvajal's account the digital display of texts heavily influenced their writing process. Carvajal notes that the ways they used the computer in the different phases of their project was determined by a combination of understanding the affordances of the computer, but also by their previous experience with academic writing, their perception of the teacher's expectations, and knowing how to participate in school.

And finally, also looking at technology at school, Irán Guerrero moves our attention from students to teachers in Mexican middle school classrooms. She examines current educational policies that promote using the computer at school through the case study of a geography teacher and how he incorporates it into his geography class. She notes that his access to technology is situated in a complex context shaped by widespread cultural, social, and political forces, but more specifically in his school's organization, classroom conditions, and school culture. She argues that, as in the case of other literacy practices, the incorporation of technology and connectivity can only be understood in the social contexts in which they are located. She explores the relation between literacy and technology by reconceptualizing what teachers recognize and understand as learning and finds that the ways teachers use technology in their work is deeply grounded in their beliefs and teaching practices.

Organizing the Book

The book is organized into three parts: Latin American Literacies: Theoretical and Methodological Approaches; Literacy and Numeracy as Social Practice: Latin American Perspectives; and Literacy and Numeracy in Education: Experiences in Latin America. There is also an Afterword by David Barton.

Part I. Latin American Literacies: Theoretical and Methodological Approaches.

This first part presents three chapters that showcase different ways of approaching the study of literacy. Marildes Marinho's chapter, "The Local and the Global in Literacy Practices in 'Traditional Communities'" (in Brazil), discusses the ongoing search for ways to conceptualize reading and writing by following the debate about terminology in Latin America and illustrating how different ways of thinking and naming literacy-related phenomena highlight what is believed and known about them. Her chapter is followed by Elisa Cragnolino's, who uses Bourdieu's notion of habitus to analyze the ideological role of literacy in rural Argentina in her chapter, entitled "Access to Written Culture as Part of the Social Reproduction Strategies of Rural Families in Córdoba (Argentina)." Writing about a different part of Latin America, Enna Carvajal in her chapter "A Multimodal Approach to the Understanding of Students' Collaborative Writing of Digital Texts" explores multimodal theory as a way of understanding how students use a computer to collaborate and produce complex texts in a high school science class in Mexico City.

Part II. Literacy and Numeracy as Social Practice: Latin American Perspectives.

Here we present chapters about literacy and numeracy practices and use in a variety of settings and complex contexts in Latin America. They share the common theme of how people's beliefs shape what they are willing to do, or not do, with literacy and numeracy. The events discussed here are located in a number of scenarios, from the city of Córdoba, Argentina, to the coast of Mexico; from households in Belo Horizonte, Brazil, to a program for youth in Santo Domingo; and to a market place in Michoacán, Mexico. They examine a variety of social interactions and contexts as well: Kalman's chapter, "GPS Technology, Map Reading, and Everyday Location Practices in a Fishing Community" looks closely at how fishers learn to use hand-held digital locating devices via marine scientists without giving up more traditional knowledge; Lorenzatti in her chapter "When Illiterate isn't Illiterate: Reading Reality in a Multimodal Way" describes in great detail how a woman with no schooling interprets different forms of representation to interact with more expert others in order to understand complex texts and situations; Maria Lucía Castanheira in "Indexical Signs within Local and Gobal Contexts: Case Studies of Changes in Literacy Practices across Generations of Working Class Families in Brazil" analyzes indexical signs as a way of identifying and exploring changes in literacy use and beliefs in cross-generational data in an urban working-class township. This chapter is followed by Irma Fuenlabrada and María Fernanda Delprato's study on indigenous forms of community relationships and their impact on numeracy practices in Mexico, presented in the chapter "Survival of Original Knowledge."

Part III. Literacy and Numeracy in Education: Experiences in Latin America.

The final set of chapters looks at literacy and numeracy in a variety of educational settings. The first chapter of this section, written by Shirley Brice Heath and Daniel Sobol, narrates how a well-planned community program in the Dominican Republic becomes unpinned when its designers fail to understand and cultural beliefs in depth. Their piece, "When Literacy Brings Too Many Risks: A Successful Lesson in Failure" shows how North American notions of development and progress are not necessarily shared by Latin Americans. Gelsa Knijnik's chapter, "The Brazilian Landless Movement and a Mathematics Education Research Program," discusses issues related to pedagogical appropriateness for teaching mathematics to traditionally unschooled populations, while Gloria Hernández examines how out-of-school literacies are taken into account or ignored in schools in rural Mexico. In her chapter, "Reading, Writing, and Experience: Literacy Practices of Young Rural Students," she is also concerned with making schooling relevant for young middle school students who are often the first in their families to complete a primary education and go on to further schooling. Situated in an urban junior high school in Mexico, Irán Guerrero examines how a teacher struggles to introduce the use of the computer and the Internet into his classroom with little guidance or support from his school system. In her chapter, "Technology and Literacy: Towards a Situated Comprehension of a Mexican Teacher's Actions," she is able to show how specific contextual conditions influence his understanding about literacy, learning, and his purposes in teaching. Elsie Rockwell, also situated in a classroom in Mexico, but this time at the elementary level, illustrates with ethnographic data different ways teachers mediate how students read and interpret official school textbooks in her chapter entitled "Preambles, Questions, and Commentaries: Teaching Genres and the Oral Mediation of Literacy." The closing piece of this collection of chapters is written by M. Sidury Christiansen and Marcia Farr, who discuss the impact of transnationalism on language ideologies in Veracruz, a state on the Gulf of Mexico. Their contribution, "Learning English in Mexico: Transnational Language Ideologies and Practices," centers on how immigration and repatriation experiences are influencing and often changing how people approach learning English, how they value it, and their expectations about what knowing English might bring them.

The rich set of accounts included in this volume makes a significant statement, we believe, not only to the research field but also to policy and practice. In questioning the dominant and narrow perspective on literacy still held in many circles, the authors demonstrate the contribution to theory and to data collection that a social practice view of literacy offers. And at the same time, by showing in such intricate detail what actually goes on in the everyday uses of reading and writing across rural and urban contexts, embracing people who migrate across these and other boundaries, and whose work lives are changing in the larger economic order, these accounts can help those responsible for policy—both in economic terms and in educational contexts—to address the complexity that such policy has to take into account. The combination of well-grounded theoretical accounts and rigorous data collection can, we believe, provide a firm basis for making such a move. That, then, is the potential contribution of this volume, and we look forward to engaging with readers from a variety of fields in extending our thinking and our practice in the area of Latin American Literacy and Numeracy Studies.

References

Barton, D., and Hamilton, M. (1998). *Local Literacies: Reading and Writing in One Community*. London: Routledge.

Barton, D., Hamilton, M., and Ivanic, R. (eds.) (2000). *Situated Literacies: Reading and Writing in Context*. London and New York: Routledge.

Blommaert, I. (2005). *Discourse: A Critical Introduction*. Cambridge: Cambridge University Press.

Collins, J., and Blot, J. (2002). *Texts, Power and Identity*. London: Routledge.

Geertz, C. (1983). *Local Knowledge: Further Essays in Interpretative Anthropology*. New York: Basic Books.

Grenfell, M., Bloome, D., Pahl, K., Rowsell, J., and Street, B. (forthcoming). *Language and Classroom Ethnography: Bridging New Literacy Studies and Bourdieu*. New York: Routledge.

Kalman, J. (1999). *Writing on the Plaza: Mediated Literacy Practices among Scribes and Clients in Mexico City*. Cresskill, NJ: Hampton Press.

Kalman, J. (2001). Everyday paperwork: literacy practices in the daily life of unschooled and underschooled women in a semi urban community in Mexico City. *Linguistics and Education* 12 (4), 367–39.

Kalman, J. (2004). *Saber lo que es la letra: una experiencia de lecto-escritura con mujeres en Mixquic [Discovering Literacy: Access Routes to Written Culture for a Group of Women in Mexico]*. Mexico, D.F.: Secretaría de Educación Pública-UIE-Siglo XXI Editores.

Kalman, J., and Street, B. (eds.) (2009). *Lectura, escritura y matemáticas como prácticas sociales. Diálogos con América Latina [Reading, Writing and Mathematics as Social Practices: Dialogues with Latin America]*. Mexico City: Siglo XXI Editores and CREFAL.

Kress, G., and Van Leeuwen, T. (2001). *Multimodal Discourse: The Modes and Media of Contemporary Communication*. London: Arnold.

Kress, G. (2002). *Literacy in the New Media Age*. London: Routledge.

Lankshear, C., and Knobel, M. (2006). *New Literacies: Everyday Practices and Classroom Learning* (2nd ed.). Maidenhead and New York: Open University Press.

Pahl, K., and Rowsell, J. (eds.) (2006). *Travel Notes from the New Literacy Studies: Case Studies of Practice*. Clevedon: Multilingual Matters.

Street, B. (1984). *Literacy in Theory and Practice*. Cambridge: Cambridge University Press.

Street, B. (ed.) (1993). *Cross-Cultural Approaches to Literacy*. Cambridge: Cambridge University Press.

Street, B. (1995). *Social Literacies: Critical Approaches to Literacy in Development, Ethnography and Education*. London and New York: Longman.

Street, B. (ed.) (2005). *Literacies Across Educational Contexts. Mediating Learning and Teaching*. Philadelphia: Caslon, Inc.

PART I

Latin American Literacies: Theoretical and Methodological Approaches

2

THE LOCAL AND THE GLOBAL IN LITERACY PRACTICES IN "TRADITIONAL COMMUNITIES"[1]

Letramento: Only a Neologism?

Marildes Marinho

The Editors of this volume, Judy Kalman and Brian Street, are sad to have to tell their readers that the author of this chapter, Marildes Marinho, suffered an unfortunate death soon after she completed the final version of the text. As readers, you will recognize when you read the chapter below that Marildes played a significant role in both research and practice in Brazil, helping to respond to policy shifts in that country in ways that supported indigenous peoples in particular, but also constructively engaged with fellow academics and with policy makers in those encounters. All of these people are shocked and saddened at her loss and numerous public displays of affection and respect were occurring as we finished our manuscript. Brian works frequently in Brazil in the department where Marildes taught and is personally shaken and upset at her loss. Judy also collaborated with her at conferences and in publications and will miss her. All of us, as fellow academics, but also as activists believing in the importance of research of the kind Marildes demonstrates here for both policy and practice, acknowledge her deep and lasting contribution. We feel sure that as you follow the narrative below, as readers you will recognize the origin of these sentiments that we are expressing here. Thank you, Marildes.

About twenty years ago, the word *letramento* (a translation of literacy) landed in Brazilian territory, in the fields of linguistics and education. Until the end of the 1980s, the word *alfabetização* (alphabetization/basic reading and writing instruction) and its correlated terms were the main words used in the Brazilian repertoire to describe the relationships that people and society had with the written word. In 2001 *letramento* appeared listed for the first time in the *Houaiss Dictionary of the Portuguese Language*, opening the doors further to related neologisms. This listing certainly did not guarantee a homogeneous and uniform meaning, just as it did not appease the controversies that the very concept would entail.[2]

In Mexico as well as in other Latin American countries literacy can be *alphabetization* or written culture and in Brazil it can be *alfabetização*, written culture, and also *letramento*. The introduction of the term *letramento* for literacy transformed the concept of *alfabetização* (alphabetization) into a mere reference in the historical process of meaning construction.

In 1995, the work of Olson and Torrance (1991), *Literacy and Orality*, was translated to Portuguese as *Cultura Escrita e Oralidade*. Then, in 2006, the title given to the classic work of Goody and Watt (1963), *The Consequences of Literacy*, was, in translation, *The Consequences of Letramento*. This Anglophone influence in our work shows that, although we are constantly dealing with the word and the concept of 'literacy', its translations were generally *cultura escrita* and *alfabetización*, and did not always correspond to the uses of the term 'literacy' in the original texts. This does not mean, obviously, that the term *letramento* has been able to fill this gap. The problem appears to be more complex: the phenomena that the respective terms—'literacy' and *letramento*—try to describe are different, because they deal with different socio-cultural contexts.

If, on the one hand, the expression *letramento* came into common use, on the other hand, its uses and meanings generated controversies. Its presence in our curricula for teacher training, lectures, and the didactic materials that we use in the teaching–learning processes of reading and writing in different disciplines—scientific literacy, historiographic literacy, digital literacy, etc.—generates an intense searching effort, from elementary school teachers, and their professors, for a more productive definition and use of this neologism. The challenge is to understand the way in which it can contribute to the improvement of pedagogical practices and to the reduction of the prevalent low levels in students' reading and writing capacities.[3]

A heated debate occurred in 2003, when the National Congress gathered in a seminar a group of researchers who argued in favor of phonological awareness as the method by which children learn to read and write.[4] One of the major outcomes of this seminar was to attribute the failure of basic reading and writing instruction (*alfabetização*) to *constructivism* and *letramento* (literacy) (Soares, 2004). This debate put pressure on Brazilian experts to discuss the declared *failure of alfabetização* within different parameters and strategies, and to emphasize a stronger connection between the process of learning writing—within the dominant system—and the social uses of the written language (literacy). In a climate of tension and dispute over models or conceptions of *alfabetização* (basic literacy), the response to the *back to phonics* movement appeal has been the "re-invention of *alfabetização*," as a way to recover its specificity, its multiple aspects, and its integration with wider literacy processes, captured for some by the term *letramento* (Soares, 2004).

Still today, we ask ourselves why we invented this neologism.[5] Among various explanations, it can be argued that the concept found fertile ground, prepared by epistemological changes in the fields of the science of language and education that provoked changes in the teaching objectives of the Portuguese language. Initiated in the 1980s, this movement is known as the "pragmatic turn" for having chosen an interactive conception of language (in contrast to a conception of language as a formal system; see Knijnik this volume). A progressive revaluation of the social uses of language was achieved by choosing the text as the teaching unit, instead of teaching the grammar of the sentence. Several works were published about the concept of literacy during the years 1980–1990: for example, *Os significados do letramento* (*The Meanings of Letramento*: Kleiman, 1995), a work that widened the circulation of the concept and motivated adherence to the assumptions of the *ideological model*, as advocated by Street (1984).

Approaches from New Literacy Studies

In that context, this conception of literacy as social practice, conveyed by the New Literacy Studies (NLS), was constantly present in our work, drawing upon socio-interactive theories that relate strongly to the socio-anthropological constraints of language and of written culture (Barton and Hamilton, 1998; Gee, 2000; Street, 1984, 2000). Another possible reason for this tendency is the fact that these studies bear a strong resemblance to the thinking of Paulo Freire, deeply marked by the political, sociological, and anthropological foundations of *alfabetização*, reading and writing. In the writings of several NLS scholars, there is a strong reason for the presence of the thought, ideology, and power relations that identify Paulo Freire's work Freire, 1985; Freire and Macedo, 1987). In these studies, as well as in Freire's work, the concept of *alfabetização* (literacy) surpasses the level of linguistic skills, proficiency in the written code, or understanding of neutral, technical skills. Instead of centering itself in such formal features, this concept focuses especially on power relations and on the policies and practices of *alfabetização*.

While this socio-anthropological conception is quite significant within the Brazilian domain of literacy work, the use of the term *letramento* (literacy) can indicate different theoretical conceptions, mainly because it focuses on specific aspects of the phenomenon in question. That diversity of perspectives allows us to assert that we are not dealing only with a question of translation; but, on the contrary, with a conceptual discussion.

Soares (1998)[6] pointed out some specificities of the term *literacy in the English-speaking countries of the first world*. Five years later, a provocative study carried out by Ribeiro (2003) also set Brazil on the search to assess *letramento*/literacy indexes, increasingly confirming the need to go beyond the assessment parameters centered on the proficiency skills of the written code, in order to construct parameters centered on the social uses of writing. In the search for reaching "literacy levels" within society, the challenge will be, then, to define what are the levels to be attained who will define them, and what objectives and interests they will have. One can also ask: would it be possible to create universal parameters for literacy (in the singular) in a society marked by profound socio-cultural differences, and, therefore, with a diversity in its practices of literacies (in the plural)? What parameters and methodologies could, for example, distinguish—without discriminating negatively—the literacy practices of people who live in cities from those who live in rural areas, or in different regions and cultural contexts?

In brief, we can say that the concept of literacy (*letramento*) is under construction. This is so, mainly, because it comprises old as well as new phenomena that are the result of changes in the ways that we relate to written culture within Brazilian society; particularly at the times of the new writing technologies (see Carvajal and Lorenzatti, this volume) and of growing pressures from social movements that are in search of a wider inclusion of marginalized groups (see Gnjinik, this volume).

However, we can acknowledge that also in Anglophone countries, and in the English language itself, the debate revolves around the same fundamental point of this discussion: what is the place for the use of the alphabetic code, and what is the place for the social uses of writing? In an interview with the journal *Lingua Escrita* [*Written Language*] in 2009, Brian Street places the debate over the use of the terms *alfabetização* and *literacy* as follows:

> The revised trend in the [debate about these terms] is not to use the term "*alfabetização*" in educational circles, but to use it to speak about historical changes or about development projects. The term "*letramento*" (literacy) is widely used to cover both references indicated by the Portuguese language with the terms *alfabetizção* and *letramento*, which means that *literacy* refers as much to learning an alphabetic code, as to the uses of reading and writing in daily life. But, the same debate can be reaffirmed; that is, there is an ongoing discussion about the emphasis that should be placed on the alphabetic code during the early stages of learning.
>
> *(Street, 88–89)*

Within that accelerated theoretical and methodological outline it also seems important to highlight that, although the literacy concept as proposed by the NLS is very much alive in Brazil, the use of its ethnographic approach in teaching as well as in research is still fragile.

Approaches from Social Semiotics

Kress (2003) calls attention to the theoretical and methodological risks and limitations that are present in the term literacy, especially when it can provoke a negative judgment about people who do not use an alphabetic system, and where letters are not their only resource for writing. The semiotician insists that "a writing system, or writing itself, is not necessarily the same as 'literacy', which is—for me—writing with letters" (61). Therefore, the term literacy is highly relevant to alphabetic cultures, but not to other cultures, for example with ideographic writing systems. Another limitation of the concept would be the fact that writing involves more than transliterating. Constantly arguing for a concept centered on the specificity of an alphabetic written language, Kress considers that the invention of the term *numeracy* is very good for referring to communication through numbers, and that this is very different from the way in which meaning is produced through the use of letters. Kress and Street (2006), in an attempt to bring together the two fields discussed here—NLS and social semiotics—agree that the extension of the term literacy to a variety of metaphorical references is unhelpful:

> For instance, in both approaches there is a worry about the stretching of the term *literacy* well beyond the NLS conception of *social practices of representation* to become a metaphor (and often much less than that) for any kind of skill or competence. One needs to ask whose interests are advanced and in what ways by the use of labels such as "palpatory literacy" (skills in body massage), "emotional literacy" (skills in affective massage?), "cultural literacy" (skills in social massage?), and so on. Of course, one clear effect of such moves is that where "a literacy" is identified, those with an interest in finding the corresponding illiterates are never far behind with their remedies. But even such uses where some aspects of literacy practices are involved—computer literacy, visual literacy—bring their own problems, not least of them the blunting of analytic and theoretical sharpness and power. Where there is a label there is already an answer; and where there is an answer, any need for questions has stopped. More significantly perhaps, there is the question of "complementarity". This poses quite simply an as yet thorny question: where does the "reach" of one theory stop—or, maybe better, begin to attenuate, "fizzle out."
>
> *(Kress and Street, 2006, ix)*

Approaches from French Literacy Studies

Whilst we acknowledge the research traditions signaled here—the NLS and social semiotics[7]—that have been influential in the academic works circulating in Brazil or through the groups that have been formed during academic exchange processes, we also need to recognize the strong influence of French authors on the conceptualization of literacy studies and written culture. That would lead us to lay out a hybrid, conceptual discussion—that would be the result of Brazilian and foreign works as well—and that would, in turn, display features and constraints of a unique nature, and also raise the question signaled by Kress and Street, "where does the 'reach' of one theory stop" or "fizzle out". From the French influence, it is necessary to rescue the sociological, anthropological, and pedagogical tendencies, as well as the influences from French cultural history, and from the discourse analysis represented mainly by the work of authors like Pierre Bourdieu, Bernard Lahire, Roger Chartier, Jean-Yves Mollier, Daniel Fabre, Bruno Latour, Michel Foucault, Anne-Marie Chartier, and Jean Hébrard, amongst others (Chartier and Hébrard, 1995). And, whether through the influence of French publications, or by establishing direct access to his work, some of Jack Goody's writings are references for those who seek to understand the consequences of writing (or of literacy) within society.

As Barton and Papen (2010, 3) suggest, during the past decades two important currents of studies about writing have developed separately "within different theoretical and disciplinary traditions": the *(New) Literacy Studies* (Britain, North America and other English-speaking countries), and the anthropological studies about writing (France). The authors stress that "with notable exceptions, francophone research on writing is virtually unknown in the English-speaking world and Anglophone researches [*sic*] are little known in the francophone world." In the Brazilian field of pedagogy, the presence of Anne-Marie Chartier during the last 20 years stands out; whether because of her work itself, or because of her perseverance in working with teams of Brazilian researchers from universities in the north and in the south of the country, in educating doctoral and post-doctoral students, and participating in inter-institutional research projects. In short, the researchers who have established and maintained contacts with the INRP (National Institute for Educational Research) and EHESS (School for Advanced Studies in the Social Sciences), among other French universities, have exercised, and will continue to exercise, a very important influence on the Brazilian intellectual production. Finally, all of Emília Ferreiro's work, and that of others who undertook research influenced by her thinking about constructivism in Latin America, also stands out. In addition to her epistemological reflections about knowledge construction, Ferreiro, against her will, was heralded as the creator of the "constructivist method" of literacy in Brazil, a phenomenon that from 1980 to 1990 has been strongly addressed both by schools and by the media.

To conclude the first part of this chapter, I would like to evoke through the French sociologist Lahire (1999) the phenomenon of *illettrisme* (functional illiteracy), invented in France at end of the 1970s. This phenomenon is treated by the author in a perspective of public rhetoric, ethics, and stigma. By inventing the word *illettrisme*, this country would also have promoted and/or constructed this social problem. The "linguistic phenomenon" of creating a neologism also shows the social construction of discourses, in this case about *illettrisme*. However, the relationship between the reality of the inequalities of access to writing, and the discourses that address it, is not evident. In the same way, the invention of *letramento* (literacy) creates the illusion of transparency, of consistency, and of theoretical and pedagogical functionality. We use the term believing that everybody knows what we are saying, without the need to explain what we are saying.

Therefore, discussing conceptions and practices of literacy, of written culture, and of *alfabetização* in Brazil means working with diversity, with differences, and with social contradictions; all of them specific characteristics of the present social situation in Brazil.

We will now turn to the ways in which such theoretical traditions can help us understand specific policy shifts in Brazil, with specific reference to the accommodation of "indigenous" peoples into university and into government-funded projects.

The Admission of "Traditional Peoples" and of Linguistic Varieties into the University

By the end of the first decade of the 21st century, in Brazil, one of the demands of the so-called "social movements" was that the "traditional peoples"[8] had to be admitted to university. According to the rhetoric of the Landless Workers' Movement (MST), they "occupy the plantation of knowledge," or, according to other voices and depending on the point of view, "invade" "our" territory, mobilizing the academics into new places and references. The second part of this chapter outlines some questions about literacy and written culture, in which the rhetoric about inclusion, as well as the amelioration and overcoming of social injustices, is the keynote of governmental policies.

In 2006, at the last stage of the course *Pedagogia da Terra*,[9] where most students were members of the LWM, we heard from a female student—a leader of the movement—a statement given with obvious satisfaction: "we are finally occupying the plantation of knowledge" in a strategic update of the main goal of the social movement that was geared towards the reclaiming [by indigenous people] of large land-holdings or plantations.

The fact that only by the end of the first decade of the 21st century[10] did public universities in Brazil open their doors to traditional peoples reminds us of what happened in the 1970s, when the country went through the democratization of elementary education. Several Brazilian public universities opened special higher-education courses, aimed at educating indigenous and rural school teachers. The "traditional groups" had put pressure to be admitted to universities and to have access to "knowledge."

Confronted with a new student profile, these universities had to seek new knowledge and strategies to understand the expectations of the "traditional peoples" as well as their particular knowledge and expectations regarding written culture. However, the categories that we use to name these peoples—native peoples, indigenous, *quilombolas*,[11] *geraizeiros*, settlers, *negros, ribeirinhos*—do not guarantee that we know them or that we understand the supposed boundaries that distinguish them.

It is within this movement of "occupation" or "invasion," of consent and resistance, and especially of this need for dialogue, that the widespread and controversial term *letramento* enters our dictionary, our discourses, and our practices. To think about such a concept of literacy means primarily to work with the notions of diversity and of cultural and linguistic differences, as well as with the perplexity of the academic world when confronted with the "invasive" linguistic variety in the halls of our universities (Lea and Street, 1998, on "academic literacies"). Linguists, socio-linguists, and experts in language teaching are questioned about how to deal with the new university students who speak a non-standard dialect. To correct or not to correct the speech of those who use the language differently: is that the question?

Within this environment—so distant and reluctant to acknowledge linguistic diversity—these students' relationship with the reading and writing of academic genres is equally tense. We, at university, have certainly never observed before such a gap and difference in the linguistic universe of teachers and students. The admission of these groups into the universe of academic literacy provides us with new, complex, and delicate situations regarding language and its uses that deserve to be analyzed within new theoretical frameworks.

The inclusion of these students at the university amid controversy broke with the assumption that to enter a high-quality public university it is first necessary to master the standard educated language that is supposedly learned at the elementary school. These social movements indicate that it is no longer possible to wait—as it was probably wished or expected—for the eradication of linguistic varieties or the acquisition of a standard dialect within those communities before giving them access to the university. They come in with their own languages and dialects, questioning and challenging us. If the linguistic variety within the Portuguese language is an issue, the process of constructing an intercultural, differentiated, and bilingual education for indigenous peoples is not less important or new. To train indigenous teachers, we need to understand the process of "revitalization" or reinvention of indigenous languages and the bilingual curricular projects of indigenous schools. We are constantly asking ourselves: how do we build a training proposal to strengthen peoples' cultural histories without promoting further the symbolic violence—in the sense given by Bourdieu (1996)—that is typical of intercultural relations of highly stratified societies? How do we avoid that our training proposals reinforce the stigma of the "unfit" or the "dispossessed"? Or that the training of "indigenous"' people takes place in different sites and with different qualifications than the "mainstream," thereby reproducing the hierarchy of social capital to which Bourdieu refers?

Everyday contact with the members of "traditional peoples" at university and within their territories has encouraged observation and analysis of aspects of speech and writing that had not yet been approached in the discussions about *alfabetização* (basic literacy), written culture, and literacy in Brazil. The issues raised below attempt to articulate the conceptions of literacy events and literacy practices (Barton and Hamilton, 1998; Heath, 1982, 1983; Street, 2000, 1988) in those communities.

For the NLS, both the articulation and the distinction between the concepts of literacy events and literacy practices have been fundamental. For Brian Street, literacy is sustained by socially constructed epistemological principles (Street, 2003, 1988). The "literacy practices, [then], refer to the broader cultural conception of particular ways of thinking about and doing reading and writing in cultural contexts" (Street, 2003, 79). Contrary to the literacy event, the practices are not directly observable, because they "involve internal processes which are frequently subconscious, as well as values, attitudes, feelings and social relationships" (see Barton and Hamilton, 1998, 6–7; Street, 1993). At the same time, the understanding of literacy practices has a discursive dimension. As stated by Barton and Hamilton, the notion of literacy practices "includes people's awareness of literacy, constructions of literacy and discourses of literacy, how people talk about and make sense of literacy" (6). The uses and the processes of construction of meaning around a text, whether in reading or in writing, are not based only on the situation of communication or on the speakers' immediate goals (literacy event); they are part of a larger scheme, the socio-historical context of the discourse, or of the conditions of discourse production. In addition to the "shared cognitions represented in ideologies and social identities" (Barton and Hamilton, 1998, 7), the meanings attributed to the texts and to literacy are produced within an inter-discursive network marked by specific socio-cultural determinants. The dialogic approach to the functioning of discourse, as well as the fundamentals of the theory of genres, as proposed by Bakhtin, have been productive in the construction of a social theory of literacy.

As stated previously, the main meaning of the "pragmatic turn" in the field of linguistic studies is adhering to a declarative conception of language, focusing on the socio-historical context (political and ideological constraints) and the immediate situation of communication (also Hymes, 1994), as well as an approach to discursive textual genres. Therefore, analyzing literacy practices and literacy events in any social space means describing the rules behind them, taking into account the interactive situation (between the individuals and their goals, or between the individuals and the object of their interaction); the written material (textual genres and their supports) and the ways of relating to this written material; the verbal interactions that reveal the negotiations of meaning; and the effects of meaning that are established around, or are part of, these texts (Marinho, 2007; Marinho and Murta, 2008). All these elements situate the literacy practices and literacy events in the socio-historical contexts of literacy practices, which in turn are embedded in the institutions that produce them, confronting them with power relations as well (Street, 2000).

New Writing Practices in Traditional Communities

In the last two decades, the Xacriabá indigenous communities from the Minas Gerais region of Brazil have undergone a process of significant changes in the way they relate to written culture, particularly for four reasons: (a) the creation and consolidation of schools for basic education; (b) the arrival of television and the computer; (c) the establishment of training courses for middle and higher education indigenous teachers;[12] (d) the demands for the design of self-managed projects within a network of several community organizations and a unified association for educational affairs that includes all the communities in the Xacriabá territory. Therefore, although the demands are external and supported by their own rules and principles, project design in the community associations is guided by the principle of collective management that is already operating in those communities; and, in turn, this collective management has generated a demand for internal communication through writing. Another factor influencing these relations with writing is the presence of the university (of students and teachers) who have been participating in the process of establishing schools, implementing research activities, and giving support to educational, economic, and cultural projects in this indigenous territory.

The group of university professors and students is often faced with the need to mediate situations related to the production of written texts that to some extent are unknown to them and completely unfamiliar to Indians. This mediation puts two groups from different cultural backgrounds in contact, particularly regarding the world of writing. In the meetings between indigenous people and people from the university in 2004/2005, it was easy to identify the latter, as they were carrying diaries, notebooks, pencils, pens, cameras, tape recorders, and video cameras. In working meetings, writing was becoming increasingly necessary, and complementary to oral practices. Xacriabá indigenous people recognize how writing has become part of their social organization, leading them to think about the changes that this practice has brought to the ways they interact. They emphasize the increase in the number of meetings, the time gaps between meetings, and the dynamics of their presence at these meetings. According to the President of the Association, the presence of new participants at each meeting could interfere with the continuity of the discussion. Thus, the minutes of the meeting summarize what was discussed, and ensure an accurate recollection of facts for those who attended it. Certainly, the presence of writing does not occur without doubts and struggles. Sometimes, when someone proposes to take notes at a meeting, a community leader—the one who is always very committed to cultivate the indigenous group's traditions—challenges him or her by saying that what counts for them is the spoken word: "*We had never had this here, what counts is our word, what is agreed is agreed.*" In contrast, another community leader—who has been involved in countless meetings—argues: "*I really admire your head [your memory], so far, you right? But I doubt it, the way it's going, you will succeed to memorize this.*"

In addition to helping them with the demands from other institutions, such as projects and other documents to be sent to ministries, and municipal and state departments, the presence of the university participants in these communities contributes to the development of new relations with the spoken and written language: distance communication by telephone, email, notes, and messages. There is also an emphasis on new genres of writing, such as the use of an agenda, on taking notes in working meetings, and on the production of written texts for community and political purposes. These texts become support objects and production processes of a shared interpretation, and a conjunct declaration that is marked by the constant search for common goals, expectations, and knowledge that are often different and differing. The events organized in partnership with the university develop new literacy practices in these communities.

Among these practices, there are projects that must be produced by community associations in order to obtain funding from the Federal Government, such as the *Programa Fome Zero.*[13] Xacriabá Indians, however, did not know exactly what this project would be, or how to write the project document. To support the execution of this project, a university group (teachers and students) developed, between 2003 and 2006, a set of actions with this population. One of these actions was to carry out an assessment of the local economy, a project for which one of the associations had already obtained funds and was in need of support to implement it. In order to achieve this goal, the teachers and students worked with representatives, leaders, and presidents of the community associations of the Xacriabá indigenous villages (Minas Gerais) in the implementation of project development and project management workshops. In the beginning this task seemed to be less complex than what we had experienced up to that moment. Apparently, it would have been enough to sit in front of a computer with the ones involved in the project and fill in the information demanded by the Ministry. The technical part, or the search for information on how to plan a garden, for example, was done by the university students, and one of them already had experience in providing advisory services in this field. Upon finishing the text, it should have been sent to the Ministry. However, even for the "most learned" within our society, the task of writing a project document and of submitting to the demands of institutional writing is incredibly difficult. And yet, we know that the success of a project depends not only (and not always) on its content. The understanding and control of the pragmatic factors that interfere with this type of work require learning and strategies that are unknown to these indigenous people.

Participants from different ministries crossed paths while working with the indigenous peoples. This often led to overlapping of interests and, more than occasionally, to disputes, abuses, and confusion. We soon realized that, between what we were experiencing and what should have been planned for the project, there was a gap that in many cases needed to be quickly understood and overcome. According to one professor, "the activity of writing things on a piece of paper is one thing, but the field they will plant is another." Between the act of writing, of describing what will be done, and planting the field itself, there are unimaginable distances and differences. The conception of time and of the way we live and organize daily life, and the way we talk about it, is different from the time that we take to plan this daily life through writing. The pattern of the complex process of construction of authorship is revealed very quickly:

> The heaviest, densest idea is that they will programmatically organize themselves to accomplish something that is on the order of the daily absolute, on the order of what is almost unspoken—to explain how it is that I plant a field—through a written document that has to be of their own authorship.
>
> *(Lia, interview, November 2005)*

A document that represented the "authentic authorship" of Xacriabá people, produced for a negotiating partner/counterpart and within a particular bureaucratic context, meant that it was difficult for them to configure the necessary set of images (Gadet and Hak, 1990) for the production of the text and for achieving the goals of such a text effectively. The fact that the mediation by the group of academics can lead to an ad hoc production of writing that is quite far from the reality of the Xacriabá became evident in the first project development workshops and during the process of applying questionnaires for the assessment of the local economy.

And it is here where the most important issue of this experience lies: understanding the place and the strategies of the university as mediator of these practices. A relevant question for the reformulation of those strategies would be: how could one line up alternatives to turn the projects' key stakeholders—Xacriabá Indians—into the projects' de facto authors? How could they appropriate such writing, recognizing themselves in the written text, and understanding the logic of that writing? How could they understand the logic of their audience–local or not–its implications, and its effects in their actions within the context of the collective management of the project? How could they conduct this process, so that they assume all the conditions and consequences of implementing the project?

It is reasonable to assume that Xacriabá people needed to be clear about the way that the contract, or the socio-communicative agreement resulting from writing, would work. In spite of the fact that it is produced collectively, the one who signs the project is the President of the Association; therefore, he or she is the institutional author, and, consequently, the one who will have to bear the responsibility for the accountability and appropriate use of the money, and for reporting this to the organization within a set period of time. If this is not done, there will be serious consequences for this person and for the association. One could ask, then, to what extent and how would it be possible to teach such knowledge, authorship, and responsibility—that is, the pragmatic logic of writing?

Beyond Writing Abilities: Building a Place of Authorship

The *literacy events* developed within these *text production workshops* raised pertinent issues for the understanding of aspects of written culture, which could not have emerged in the daily practices of these groups. They allowed us to think about the ways in which the act of writing creates and establishes significant aspects of the identities of these and other groups. One of the aspects deals with the need for greater questioning of the social practices of these subjects, especially as they learn from the oral practices, skills, and the competencies specific to writing.

Subjects considered to be barely literate, or illiterate, identify within their own experience, specific pragmatic functions, and the power of writing, which allows them to develop strategies to respond to a situation in which writing is the legitimate procedure. Another aspect is the possibility of understanding the place of authorship, at a time when we observe that there is a strong discussion about reading communities. Since writing is a constitutive element of a personal and a collective identity (Fabre, 1997, 20), we find ourselves asking: how has writing been happening in these traditional communities at a particular moment in the history of these people?

The literacy events evident in the contexts described above represent unique situations in relation to writing, showing us what abilities, literate tendencies, and power relations are involved in literacy practices that are the result of writing production demands from governmental institutions external and distant from the daily lives of these indigenous people. Since the demands for written documents were foreign to them, Xacriabá established a new dialogue with speakers, objectives and strategies that were clearly different from the ones that take place in their daily lives. For instance, I would emphasize the complexity of the tax writing process that is imposed upon these Indians, by the superposition of interlocutors at the time of production of these texts: on the one hand, the mediators of this writing (university teachers and students); on the other hand, the intended reader, the Ministry. The place of the pragmatic language skills, in which the borders between speech and writing frequently become less visible, becomes evident. This also means that the people who master the written code do not always have the best knowledge or the best declared strategies for the production of these texts. So, where and how do "barely literate", "semi-literate," and "illiterate" subjects acquire such abilities and skills?

It would be naïve on our part not to consider the interaction that takes place within the relationship with the mediators of this writing. Would the indigenous people be passing this task on to the university students out of total ignorance; or because by taking part in the production of writing, we "writers" could somehow also assume the consequences of this writing and the responsibilities that are established by writing and for writing? Or would our place as "writers" also constitute a bartering unit, like the ones that have always been used for exchanges between the "natives" and the "civilized," since the university also has specific interests such as research, a space to educate our students, etc.? In several community meetings (including the ones planned by the indigenous people), the university students were strongly pressured to take on that writing role, to take notes, summarize, or write reports, de facto preventing the indigenous participants from taking over the writing process, and setting up a potential trap within that interaction that needs to be discussed. An excerpt from a university student's field diary gives us the dimension of this awkward situation. It refers to a meeting of indigenous people in which they discussed a document from the Ministry of Education and Culture (MEC) containing proposals for indigenous education. The outcome of this discussion was going to be presented by their representatives at a Conference on Indigenous Education:

> At the end of the discussion, the group had to present a written report of what had been discussed, as well as the proposals and demands to be taken to the Conference. Initially, the women complained about not having understood what needed to be done. My first reaction was not to interfere in the discussion and to only place myself in the role of observer—something that, according to my experience, did not seem possible. But, the situation was getting very uncomfortable because they could not articulate their ideas—and, most likely because of my presence, that brought great discomfort to me and to them. And someone spoke at the meeting, the other participants, complained not "being able to put on paper" what was being said. It was after I felt terribly embarrassed by [the way] they were looking at me and at my notebook—as if I was the one or the personification of the one who mastered the activities that were difficult for them, such as organizing their thinking and putting it into writing—that I offered to help. **Then, one of them said, quite bluntly, that it was indeed good that I helped because, after all, the Xacriabá want help and not just "people to come and take notes of everything we say." In general, they seemed to like the idea and handed me the MEC document and the piece of paper on which they were supposed to take notes for me to read and write what, in my view, was being said. My intention was only to assist with the writing, putting the words on paper, and not to determine what should be written.** When I explained this to them, I felt an air of disappointment and even of contempt. But at the same time, at some moments I took on full responsibility for writing the report.
>
> *(Excerpt from a university student, 2010)*

This and other situations also encourage us to think about the role of the literate groups within these communities; whether they are local, external, or NGO representatives, professors, university students, or any other mediator of literacy practices. How does one establish and define the roles of these "mediators"? What happens in this dialogue between academics and indigenous people?

In the context of the external demands from governmental institutions, could one ask: to what extent do such indigenous people make use of the legal writing that is destined for the institutions that organize society? The attendance sheet would have the purpose of proving they had taught a class, a condition for getting paid. What other mechanisms, strategies, and materials would fulfill legal purposes in their social practices? Since we know about the existence of these purposes in other social and historical contexts in which writing was not present, we also need to assume that they exist. They have their rhythm, their timing, and their own way of establishing commitment, and of making deals and contracts that do not necessarily pass through a written record.

There is a collection of writing practices in these communities; their primary function is to respond to the demands of governmental institutions. However, not complying with these demands can generate serious consequences for the indigenous people. In both of the situations cited above (writing a project, the attendance sheet) there is a conflict within the demands of governmental institutions for written replies. Failure to comply with these demands can generate serious effects for the government's stated partners: the indigenous peoples. They find themselves confronting oral as well as written operation models, which demonstrate the alternative and complementary character of one in relation to the other. The underlying knowledge of the world of linguistics that Xacriabá have is updated in a context of articulation or of interaction between speech and writing. As Heath (1982) reminds us, speech events can repeat, reinforce, amplify, adjust, or contradict what is written. Therefore, it is necessary for the subjects of the interaction to learn how to identify the moment when the written takes precedence over the oral. This connection between speech and writing is an important principle, highlighted by New Literacy Studies researchers:

> They do not see literacy as progress, in relation to the oral practices of language, but as complementary to them. Contextualized literacies are situated in the spheres of daily life in which writing is present, but do not surpass or have priority over the oral; we can say that oral language is the context for written language.
>
> *(Collins and Blot, 2003, 37)*

In conclusion, it must be highlighted that some obstacles, generated by the difficulty of the indigenous people to meet certain demands of use for writing official documents, have led to the hypothesis of substituting the written text (printed or handwritten) with audiovisual records—for example, the CD-ROM. (In fact, meetings have been repeatedly registered on video.) However, it seems wise to think about the political factors related to the literacy practices and models with which these communities are confronted. It would be hard to imagine that institutions that are organized according to the logic of writing would have the conditions to deal with a video as a legal document. Another complicating element concerns the power relations, or the discriminatory effects that could be projected upon these communities for not knowing how to deal with that type of official writing. The substitution of videos for written text could be understood as a certificate of "illiteracy" or of "functional illiteracy," generating a stigmatized image of indigenous peoples and creating reluctance among the agents from the governmental institutions who are responsible for the projects and processes of interest to these communities. And, as a local leader points out, "writing must prevail in this situation; besides, video technology is not always available, and paper is."

The external demands for writing (minutes, attendance records, reports, project documents, etc.) encourage us also to think about the introduction into these communities of genres that until then were nonexistent and unnecessary to their cultural practices. The need to "register in text" for later interpretation happens as much in oral cultures as in written ones, but the demand to do this through writing is a circumstance of specific socio-cultural contexts (Feldman, 1991). We can already observe that these communities are beginning to develop strategies to deal with these recently created demands. There are specific processes and means of appropriating new technologies; new discursive genres; and a developing connection between "tradition" and "innovation," oral and written, and local and global.

The interaction between university students and academics with indigenous populations has led us to reflect on the role of the literate groups that are present in those communities; whether the local groups that learned literacy through these interactions or external groups from universities. These groups are the mediators of the new literacy practices: therefore, as we intend to establish an intercultural dialogue and to set up mechanisms for the inclusion of these individuals in economic, educational, and cultural management practices, it is important to understand the way in which we construct the role of mediators in such a way that it will meet the expectations and interests of these indigenous people.

Notes

1 Research financed by CAPES e FAPEMIG.

2 For recent discussions on this same topic, Marinho (2009).

3 "Within a group of 14.1 million people, the illiteracy rate for those who were 15 years old or older fell from 13.3% in 1999 to 9.7% in 2009." Applying as criterion three years of formal schooling, around 20 percent of this population was functionally illiterate. (IBGE [Brazilian Institute of Geography and Statistics], Synthesis of Social Indicators, 2010; page accessed on 03/03/2011.)

4 This debate occurred in several countries; for example, between proponents of whole language and phonics-based approaches in the U.S.A.; in France, in March 2006, the Minister of Education Gilles de Robien by means of a ministerial regulation prohibited the use of the global method and demanded that teachers adopted the syllabic method. A teacher (Roland Goigoux) from the Ecole Supérieure d'Education Nationale (Higher School of National Education) was expelled for opposing the minister's position.

5 In the Portuguese from Portugal, the translation of *literacy* is *literacia*.

6 In 1988, the author was calling attention to the fact that studies about the social function of writing were almost non-existent in Brazil, while "in the international literature, although recent, they were already numerous."

7 Barton and Hamilton, 1998; Collins, 1995; Heath, 1983; Street, 1984; Kalman, 1999; Gee, 2000.

8 *Traditional populations* were recognized by Presidential Decree no. 6040 of 7 February, 2007. "Traditional Peoples and Communities are: culturally differentiated groups

that are recognized as such, that possess their own ways of social organization, that occupy and use territories and natural resources as a condition of their cultural, social, religious, ancestral and economic reproduction, utilizing knowledge, innovations, and practices generated and transmitted by tradition."

9 Preparatory course for teachers in rural schools.

10 In 2008 the government created REUNI—Restructuring and Expansion Program for Federal Universities—as part of the democratization process for a free and high-quality higher education for excluded sectors, as is the case of the "traditional peoples and communities" (indigenous and rural).

11 *Quilombolas* (residents of *quilombos* in Brazil, descendants of slaves who escaped from plantations that existed in Brazil), *geraizeiros* (traditional peoples of the Cerrado, living in rural areas in the northern part of Minas Gerais; the name derives from the regions of the Cerrado known as "Gerais"), *settlers*, *negros*, *ribeirinhos* (traditional peoples who live near the rivers in various parts of Brazil; generally, they fish, plant, and extract natural materials for their own use) do not guarantee our knowledge of those cultures nor the supposed boundaries that distinguish them.

12 In 1995, a process of teacher training was started, and until 2003 100 teachers graduated. Today most of them are registered in a university course at UFMG. Since 2008, UFMG, through REUNI, allows, via a college entrance exam, the registration of 35 members of indigenous communities in the Licenciatura Intercultural Indígena (BA in Intercultural Indigenous Studies), and in 2010 twelve special vacancies were opened for them in other degree courses.

13 "**Zero Hunger** is a strategy promoted by the federal government [since 2003] to guarantee the human right of adequate meals for people with limited access to food. Such strategy falls within the promotion of food and nutritional security, social inclusion, and citizen rights of the people who are most vulnerable to hunger" (http://www.fomezero.gov.br/). The proposal for *Zero Hunger* for indigenous populations is linked to the Ministry of the Environment (MMA). The project, developed by UFMG with funding from the Ministry of Agricultural Development, was proposed by the Secretary for Family Farming (attached to this Ministry), with financing that took into account the specificities of indigenous populations.

References

Bakhtin, M. (1981). *The Dialogic Imagination.* Austin, Texas: University of Texas Press.

Barton, D., and Hamilton, M. (1998). *Local Literacies: Reading and Writing in One Community.* London: Routledge.

Barton, D., and Papen, U. (eds.) (2010). What is the "anthropology of writing"? In D. Barton and U. Papen (eds.), *The Anthropology of Writing: Understanding Textually Mediated Worlds.* London: Continuum.

Bourdieu, P. (1996). *Ce que parler veut dire: l'economie des échanges linguistiques.* [*The Meaning of Speech: The Economy of Linguistic Exchanges*]. Paris: Fayard.

Chartier, A., and Hébrard, J. (1995). *Discursos sobre a leitura—1880–1980* [*Discourses about Reading—1880–1980*], Alberto Luis Bixio, trans. São Paulo: Ática.

Collins, J. (1995). Literacy and Literacies. *Annual Review of Anthropology* 24, 75–93.

Collins, J., and Blot, R. (2003). Situated approaches to the literacy debate. In J. Collins and R. Blot (eds.), *Literacy and Literacies: Texts, Power and Identity.* New York: Cambridge University Press.

Fabre, D. (1997). *Par écrit: ethnologie des écritures quotidiennes* [*In Writing: Ethnology of Everyday Writing*]. Paris: Éditions de la Maison des Sciences de í Homme.

Feldman, F. C. (1991). Metalinguagem Oral [Oral Metalinguistics]. In D. Olson and N. Torrence (eds.), *Cultura escrita e oralidade* [*Literacy and Orality*], Valter Lellis Siqueira, trans. São Paulo: Ática, 55–74.

Freire, P. (1985). *The Politics of Education: Culture, Power and Liberation.* NY: Bergin & Garvey.

Freire, P., and Macedo, D. (1987). Literacy and Critical Pedagogy. In P Freire and D. Macedo (eds.), *Literacy: Reading the Word and the World.* London: Routledge and Kegan Paul, 141–59.

Gadet, F., and Hak, T. (1990). *Por uma análise automática do discurso: uma introdução à obra de Michel Pêcheux* [*Towards an Automatic Analysis of Speech: An Introduction to Michel Pêcheux's Work*]. Campinas: Ed. UNICAMP.

Gee, J. (2000). The New Literacy Studies: from "socially situated" to the work of the social. In D. Barton, M. Hamilton, and R. Ivanic (eds.), *Situated Literacies: Reading and Writing in Context.* London: Routledge, 180–96.

Goody, J., and Watt, I. (1963). The consequences ofliteracy. *Comparative Studies in Society and History* 5 (3), 304–45.

Heath, S. (1982). Protean shapes in literacy events: ever-shifting oral and literate traditions. In D. Tannen (ed.), *Spoken and Written Language: Exploring Orality and Literacy.* Norwood, NJ: Ablex, 91–117.

Heath, S. (1983). *Ways with Words.* Cambridge: Cambridge University Press.

Houaiss, A., and Villar, M. S. (2001). *Dicionário Houaiss da Língua Portuguesa* [*Houaiss Dictionary of the Portuguese Language*]. São Paulo: Objetiva.

Hymes, D. (1994). Towards ethnographies of communication. In J. Maybin (ed.), *Language and Literacy in Social Practice.* London: Open University Press, 11–22.

Kalman, J. (1999). *Writing on the Plaza: Mediated Literacy Practices among Scribes and Clients in Mexico City.* Cresskill, NJ: Hampton Press.

Kleiman, A. B. (1995). *Os significados do letramento: uma nova perspectiva sobre a prática social da escrita* [*The Meanings of Letramento: A New Perspective on the Social Practice of Writing*]. Campinas: Mercado de Letras.

Kress, G. (2003). *Literacy in the New Media Age.* London and New York: Routledge.

Kress, G., and Street, B. (2006). Multi-Modality and Literacy Practices, Foreword to *Travel Notes from the New Literacy Studies: Case Studies of Practice.* In K. Pahl and J. Rowsell (eds.), *Travel Notes from the New Literacy Studies: Case Studies of Practice.* Clevedon: Multilingual Matters, vii–x.

Lahire, B. (1999). *L'invention de "l' illettrisme": Rhétorique publique, éthique et stigmates* [*The Invention of "Analphabetism": Public Rhetoric, Ethics and Stigmas*]. Paris: La Découverte.

Latour, B. and Wooglar, S. (1986). *Laboratory Life. The Construction of Scientific Facts.* Princeton, New Jersey: Princeton University Press.

Lea, M., and Street, B. (eds.) (1998). Student writing and faculty feedback in higher education: an academic literacies approach. *Studies in Higher Education* 23 (2), 157–72.

Marinho, M. (2007). Que novidades trouxeram os novos estudos sobre o letramento? [What innovations on literacy did the new studies bring?]. In *VIII Encontro de Pesquisa em Educação da Região Sudeste.* [*VIII Southeast Region Research Encounter on Education*]. Vitória: UFES, 1–14.

Marinho, M. (2009). Nuevas alfabetizaciones en los procesos sociales de inclusión y exclusión [New literacies in the social processes of inclusion and exclusion]. In J. Kalman and B. Street (eds.), *Lectura, escritura y matemáticas como prácticas sociales: Diálogos con América Latina.* [*Reading, Writing and Mathematics as Social Practices: Dialogues with Latin America*]. Mexico City: Siglo XXI Editores and CREFAL, 40–43.

Marinho, M., and Murta, F. S. (2008). Gênero jornalístico ou parábola do bom ladrão? (O que nos revela um evento de letramento na sala de aula) [A journalistic genre or the parable of the good thief? (What a literacy event in the classroom reveals to us). Paper presented at the *Encontro Nacional de Didática e Prática de Ensino 14 (ENDIPE)*. [National Meeting on Didactic and Teaching Practices 14 (ENDIPE)]. Porto Alegre: PUCRS.

Olson, D., and Torrance, N. (1991). *Literacy and Orality*. Cambridge: Cambridge University Press.

Ribeiro, V. M. (2003). *Letramento no Brasil* [*Literacy in Brazil*]. São Paulo: Global.

Soares, M. B. (1998). *Letramento: um tema em três gêneros* [*Literacy: A Theme in Three Genres*]. Belo Horizonte: Autêntica.

Soares, M. B. (2004). Letramento e alfabetização: as muitas facetas [Letramento and literacy: the multiple facets]. *Revista Brasileira de Educação* 25 (Jan/Feb/Mar/Apr), 5–17. Retrieved 2 March 2011 http://www.scielo.br/pdf/rbedu/n25/n25a01.pdf.

Street, B. (1984). *Literacy in Theory and Practice*. Cambridge: Cambridge University Press.

Street, B. (1988). Literacy practices and literacy myths. In R. Saljo (ed.), *The Written Word: Studies in Literate Thought and Action*. New York: Springer-Verlag Press, 59–72

Street, B. (1993). (ed.) *Cross-Cultural Approaches to Literacy*. Cambridge: Cambridge University Press.

Street, B. (2000). Literacy Events and Literacy Practices. In M. Martin-Jones, and K. Jones (eds.), *Multilingual Literacies: Comparative Perspectives on Research and Practice*. Amsterdam: John Benjamin's, 17–29.

Street, B. (2003, July). The limits of the local—"autonomous" or "disembedding"? Paper presented at the Tenth International Literacy and Education Research Network Conference on Learning, Institute of Education, University of London, London.

Street, B. (2009). Entrevista com Brian Street [Interview with Brian Street] [Electronic Version]. Língua Escrita. [Written Language]. Retrieved 2 March 2011 http://www.ceale.fae.ufmg.br/linguaescrita/artigo.php?id=1&pg=2

3

ACCESS TO WRITTEN CULTURE AS PART OF THE SOCIAL REPRODUCTION STRATEGIES OF RURAL FAMILIES IN CÓRDOBA (ARGENTINA)

Elisa Cragnolino

Since the advent of New Literacy Studies (NLS),[1] the plurality of practices of written culture has been recognized and it is clear that an understanding of these is possible only if one considers the particular context in which they develop. In the account provided in this chapter, then, literacy is conceived as historically and culturally situated, constructed, and embedded in power relations. Ethnographic methods help to study the particular forms and meanings that these assume, in order to explain, among other issues, how they provide, or otherwise, access to resources and opportunities. The aim here is to use this approach to study the way in which social conditions and relations are linked with what people do with reading and writing in practice amongst rural families in Córdoba (Argentina).

In a previous study (Cragnolino, 2009), given the concerns raised by NLS, I discussed the analytical possibilities involved in putting into play the perspective of the French sociologist Pierre Bourdieu as a way to understand the practices and processes of access to and appropriation of written culture.[2] Here, I am interested in taking up some of these ideas again, to go more deeply into how certain practices of reading and writing are incorporated and transformed as part of the social reproduction strategies of families of peasant origin.

I reconstruct the history of a group of rural families living in the north of Córdoba, Argentina, for over 40 years, between 1930 and the mid 1970s. My aim is to analyze the changes in the relationships they establish with reading and writing as their objective and subjective positions change in the economic, social, and cultural fields.

Contributions from the Perspective of Pierre Bourdieu for Understanding Processes of Access and Appropriation of Written Culture in Peasant Families

Taking the perspective of Pierre Bourdieu (1988a) implies recognizing that practices of written culture and people's perceptions about them are configured in the encounter between the objective possibilities offered by different social fields, and the socially differentiated dispositions, the habitus, which, associated to specific types of capital, define the interest for those possibilities.

Bourdieu understands by field a system of positions and relationships between positions, and by habitus a system of internalized structures, subjective but not individual, that structures practices and representations (Bourdieu, 1988a). The field is a social arena of struggle over the appropriation of certain species of capital and the author identifies various types of capitals and fields; the principals are economic, social, cultural, and symbolic (see also Grenfell et al., forthcoming, for an attempt to link Bourdieu's concepts to those in literacy study).

Habitus is the social-made body. Bourdieu mentions the

> system of durable, transportable dispositions, structured structures predisposed to function as structuring structures, that is, as principles which generate and organize practices and representations than can be objectively adapted to their outcomes without presupposing a conscious aiming at ends or and express mastery of the operations necessary in order to attain them. Objectively "regulated" and "regular" without being in any way the product of the obedience to rules, they can be collectively orchestrated without being the product of the organizing action of a conductor.
>
> *(Bourdieu, 1991, 88–89)*

I find this a fertile analytical approach which, based on concepts such as field, capital, habitus, disposition, interest, strategy, and history, aims dialectically to unite structure and institutions to the action of the actors. It enables me to understand the processes by which the peasant families that are the subjects of my studies define and organize their everyday lives, develop their projects, and include education and written culture among their strategies.

It is in this line that I propose that practices of written culture may be considered as part of the social reproduction strategies, understood as that set of practices by which individuals and families tend, consciously or unconsciously, to conserve or improve their position in the structure of class relationships (Bourdieu, 1988b, 122). In this perspective, social reproduction strategies depend on the "state of the system of instruments of reproduction," which vary according to different historical periods and contextual spaces. This means thinking not only how the school system and educational provision configure objective opportunities for access to written culture, but also taking into account how the production structure, labor market, probate law, the law regulating land ownership, social policies, social security systems, etc., do this.

The strategies also depend on the differential performance that these instruments of reproduction can offer families according to their objective position, i.e. to the volume (quantity) and structure (conformation) of their resources, the economic, cultural, and social capital at their disposal. There are some resources available that ease access to written culture: there are objectified cultural capital (libraries, books, magazines, but also other reading materials related to domestic, religious, community life); there are also embedded resources (knowledge, information, skills), which are important to identify. It is important for analytical purposes to recognize also how economic and social capital facilitate or otherwise provide access to these cultural resources.

Finally, the strategies are also configured according to the system of dispositions, the habitus incorporated throughout the process of class socialization, in this case peasant class, which lead them to think, to perceive, to define "what is possible" and "what is impossible," "what is thinkable" and "what is unthinkable," "what is for us" and "what is not for us." In relation to literacy, the aim is to identify the system of representations which frame and give meaning to the use of the resources of written culture. The habitus constitutes principles of the generation and perception of practices, and helps explain how and to what extent families implement certain alternatives of action in relation to written culture and discard others, put certain mechanisms into effect and make efforts to employ them.

But the strategies related to literacy are defined not only on the basis of the volume and structure of the types of capital considered at the time and of the possibilities that these secure for the families. They are also made according to their evolution over time, i.e. to their social history and the dispositions, also in history, which are constituted in the long-term relationship with some objective structure of possibilities.

Peasant Families, Structural Transformations, Social Reproduction and Strategies in Tulumba (1930–1970)

The families who are the subject of this analysis come from the Tulumba Department, a region of the province of Córdoba, Argentina, the agro-ecological characteristics of which (semi-mountainous, semi-arid, with an environment degraded by deforesting), since the early 20th century, present a contrast with the rich Pampa south. This is a mostly rural peasant area that has historically had the highest poverty and illiteracy rates. The peasant families cultivated small plots in the low valleys, raised livestock (cattle and goats) in an extensive system, and worked in the bush, cutting timber and making charcoal. Their members sold their labor seasonally in the southern agricultural regions of the provinces of Córdoba, Santa Fe, and Buenos Aires, or permanently in some of the ranches of the area, where they worked as herdsmen or laborers of large establishments engaged in cattle-farming.

As from the 1940s, the possibilities available to rural families for autonomous reproduction became more limited, while, in the context of welfare state policies, job and settlement opportunities in the cities increased.[3] Definitive migration to urban destinations also increased, and with it a process of depopulation of the villages and depeasantization that has still not stopped. This process is parallel to the expansion of large capital-intensive farms advancing over the production of small properties, fencing in farms and creating a new economic and social scenario.

These structural changes, briefly presented here, meant changes in urban family economic reproduction strategies. In the first decades of the last century, the families ensured their direct economic reproduction through a variety of farm activities (maize, horticulture, cattle- or goat-raising, vegetables, the bush, picking fruit, etc.), but steadily lost their capacity for autonomous sustenance as capitalism penetrated the area through the markets of land, labor, products, and consumption.

The diversification of activities, the intensification of the use of family labor, and the distribution of tasks by gender and age were the main strategies. Migration strategies then began to become ever more important. This was in the context of the high availability of labor that the families have and linked to the limited possibilities of further dividing land among the children, without affecting the economic viability of the unit. It was also related to the intention of seeking alternative resources, especially as the monetization of the rural economy increased, with the arrival of services and the purchase of manufactured products which replaced domestic production.

The reproductive strategies of the peasant units of our informants were affected during this period by the changes mentioned above and especially by labor market dynamics: on the one hand, the modification of the annual cycle occupying rural workers in the Pampa area, with the decrease in the need for manual workers, resulting from the mechanization of agricultural tasks that began in 1940 and accelerated in the 1960s; on the other hand, the development of an urban job market, especially industrial and for services (particularly in building) that turned cities, including Córdoba, into a major magnet for rural dwellers.

With the retraction and then without the contribution of resources from the work in the Pampas, with the growing deterioration of environmental conditions and of the competitive advantages of agricultural products from marginal areas such as northern Córdoba, economic reproduction in these families could only be guaranteed to the extent that some of its members left the farm and tried out labor strategies involving their definitive departure from the home. Also, greater market integration, produced by the development of means of communication that eased access to markets (both for selling the production and to buy products previously obtained through family self-supply), placed them in a situation of increasing weakness and reinforced the need to increase income.

Migration, whether seasonal or then permanent, of many of the family members introduced changes in the objective conditions of life and restricted its future as a productive unit. But it was also presented as a structuring process and affected different dimensions of the household unit. As we will see, one of these transformations had to do with the meaning and scope that was starting to be given to schooling as preparation for migration, and the relationship these changes have with written culture.

The implementation of reproduction strategies, whether related to self-sufficiency on the farm, rural or then urban migration, rural wage-earning, or inheriting the farm and the animals, put into play, in turn, educational strategies that guided different members of the unit, depending on whether they were leaving the farm, becoming rural laborers or herdsmen, or remained in charge of the family farm. The family prepared its children for these alternatives and incorporated informal educational mechanisms, and gradually later the school, into this training.

Access to School: Reading and Writing as Part of the Strategies of Social Reproduction

In Tulumba, early in the 20th century, there were primary schools (provincial and national). The sources consulted[4] show, however, that only 50 percent of children complied with compulsory education in the 1930s and that regular attendance was limited to half of those registered. The schools, at that time, attracted mainly the children of the owners of the largest farms and cattle ranches in the area, who had worked to set up the schools and provided premises for them to operate. These families thus ensured the existence of a school in a location near their homes and facilitated their children's access to education and thus to some knowledge which, in other social sectors and especially in the city, was already given social importance: to know how to read and write and to do some arithmetic.

For the rest of the peasant families, the possibilities of going to school were much more limited; there were geographical distances to cover (the schools, still not very numerous in the area, were in some cases several kilometers away), but there were also urgent material needs, which involved the children, that had to be satisfied. The inclusion of children in domestic and peri-domestic tasks began at about the age of six, and, since its intensity depended on the structure and availability of family labor, it was part of daily routines. In an interview, one of the family members in the research stated:

> Tina: [. . .] As a little child and then afterwards, everybody out to work ... Just getting used to working the land, with the animals ... country people, it seemed to them that school wasn't necessary ... and so it was that we grew up ... Nothing of learning letters; why should we? We were just in the countryside.[5]

The children were subjected to early learning about domestic and farm work through observation and then imitation, allowing them to appropriate gradually the repertoire of specialized knowledge that would serve to solve "practical" problems (knowledge of weather cycles, the operations necessary for cultivation and harvesting, handling the tools, the diseases of the animals, the routines of the kitchen, and household chores). School for these children was far away, not only physically, but socially. Education and schooling for these families did not appear as a priority need for generational reproduction.

From the 1930s, an expansion of basic education services was seen in the department of Tulumba. Between 1930 and 1970, the number of schools increased from 33 to 60, the number of teachers grew from 46 to 154, and enrollment went from 1562 to 2886 students. This situation corresponds to the strengthening of the education system as part of national and provincial state policies; but it was the result not only of state initiatives but also of new social configurations, and demands and expectations that peasant families began to have regarding the education of their children.

As mentioned earlier, the context was that of a declining population because of permanent emigration, generally to urban destinations, reduction of farms and herds, monetization of transactions, and the loss of economic sustainability of farming units. These transformations in socio-economic and educational fields influenced the manner and the efforts that families gradually applied to formal education and specific moves to approach the school. The dispositions (in Bourdieu's terms objectively adapted, structured structures predisposed to function as structuring structures, which organize practices and representations) were slowly changing and helped build new ideas about how one passed through childhood, what the obligations of children were, the children's possible futures, and the value that reading and writing would have in a future which, in most cases, involved abandoning rural life.

From the late 1940s and early 1950s, the number of children of the families being studied who attended schools gradually increased, though family efforts concerning education were still directed to ensure continuity in school only for some of the children and for a limited period, until the time when they had minimal knowledge of reading and writing and arithmetic, which would enable them to cope better when trying to find work outside the farm.

It began to be important then to "know how to sign [their name]", "to know how to read and to add" so as not to be cheated, for example, by the contractors who took them to the harvest, by the bosses in the Pampa, the mistress in domestic service, or the owners of the general stores, and the traveling salesmen who arrived in the area and provided merchandise. Recognizing their own inability to convey this knowledge, the family began to delegate these formative activities to the teachers. Another interviewee stated:

> Rolo: When the boss [contractor for the harvest] arrived, some just by word, but others threw the papers in front of you and the old ones looked at it and

> didn't know what they were signing ... You had to know, that is, someone had to know, so that they didn't put one over you, so then if the kids could understand, if they were smart, that was a great help, you understand? So then they always took care that one of the bigger kids should know ... The same with the merchandise and all ... When they [the traveling salesmen] came and they said to you, "Give me so much goat, and I'll give you the sugar, mate tea, spaghetti," they knew how to take advantage ...
>
> I was sent ... I must have been like nine [years old], but I didn't like it, so not much, to sign, I sign cute, you wouldn't believe; don't read though, sums, more for sums. But my cousin Toto, he had a head for it, so it was more reading, writing, everything—they always used to consult him ...[6]

Until then, all the job options seen as possible had been manual and did not require additional knowledge or school skills, so the children stayed in school only for the time needed to acquire this initial knowledge, as long as they were not required for farm jobs, or while they could balance school activity and work. But then the migration strategies gradually increased, which involved a greater number of domestic group members, and these had urban destinations. That was when school attendance and literacy began to occupy a bigger place in family economic and social reproduction. It was necessary to prepare children for migration to urban centers and there obtain other kinds of jobs. The school could prepare them for this migration. As Tina states:

> Tina: We started to send [the children to school] because you already knew that they couldn't be like us and that there was a lot of work, so much work in the city ... In Peron's time ... so many people went ... lots of possibilities, lots of jobs ... and there they had to know, didn't they? About letters, writing, reading. It couldn't be like we were, that here, if you don't know, well, for the goats and animals, but there, if you don't know, the minimum of writing, that they give you a paper and you don't know, or for building or I don't know what, a factory, that the bosses came and passed you a paper and if you didn't know, so many chances were lost ...[7]

Schooling was expected to provide children with instrumental knowledge to make better use of the opportunities that opened up to them in the urban space; they should acquire skills and abilities that would enable them to improve their lives and work. Reading and writing were essential aspects of those skills. As Rosa stated:

> Rosa: They were country people, just that, so ignorant they served for fieldwork and nothing else. In my house, no books or anything, but my dad did have one thing: he said we had to learn to read and write, and so he sent us, so they wouldn't look down on us, to have a future, like, if the girl went

> [to the city] or the boy, so they knew enough to look after themselves, you understand?; so they would be respected, because if you can't read, nobody respects you, you see? And so it was that one after another we were sent.[8]

Although parents did not know what kind of specific knowledge was involved in "managing letters," they recognized its importance. Reading and writing would not only open the way to "decent jobs" and better-paid ones, but it began increasingly to be thought of as a means of social empowerment; the ability to read and write opened up the "respect" of others, as Rosa said: "It makes its bearer visible and in a way protects them and puts them in better standing".

As the years went by, the housekeeping needs of some of the study units, particularly those with more economic capital, started to include some resources related to school attendance. On the one hand, there were material needs to be resolved, they had to buy supplies and books, but they also had to ensure that the children had enough time to go to school, and space and time to do their homework. Mercedes states:

> My dad came home late, tired of laboring like a donkey all day ... but he called us and made us show him our school books ... I don't know what he saw because he didn't know. He had gone to school, but the lack of practice, he didn't know anything, but he looked at them the same ... José, who skipped off more often than not [from school] and stayed around with a friend killing birds or doing I don't know what, had to come back and on the way write something to show the old man [laughs]. We had to show him the book, always. [9]

As mentioned by our informants, one of the greatest difficulties the children had in school had to do with reading and writing. They note here the absence of written materials in their homes, the inexistence not only of their parents' "studies" but also the "lack of practice" and the limited opportunities to make use of written language in their everyday and working lives, which distanced most members of the families from these skills.

This lack of knowledge meant that many of these parents were unable to exercise any direct "monitoring" of homework because they did not know what the children were studying. This did not mean that they always ignored these "tasks," and they sought to exercise, in any case, some control above all in terms of habits and responsibilities. As Gladis said:

> At two in the afternoon, Mom sat us down [the daughters] ... at the big table, each of us with our homework. She didn't know many things, but she got herself a dictionary and found her way around with that. She was very demanding with spelling and neatness and insisted and insisted on homework.[10]

In Gladis's family, the monitoring function involved special times in the household day devoted to this activity, some material resources (books, dictionaries), and the habit of reproducing school writing practices, which were not found in the other study units.

The Presence of Written Culture in Family and Community Life, Work and Migration

The first image that informants usually present about their rural communities and households is that of an "illiterate" environment which directly corresponds to "poverty," "lack of understanding," "ignorance," and "the work of ignorant people." However, throughout the interviews and life histories, the existence of materials of written culture and of the inclusion of family members in literate events and practices are mentioned many times.[11]

In the community and around an institution such as the school, multiple literacy practices were deployed. I am referring not only to the variety of written materials found there (posters, files, books, brochures, etc.) or to what happened in the classroom or at home about homework, exercise books and books, but to another series of activities that involved the existence of a school. There were, for example, events concerning the production of documents relating to the petition for the opening of a school. In the hamlets, committees were organized, investigations were begun into how to obtain an official opening, and they communicated with the authorities of the Ministry of Education through a series of documents.

Although, in general, these procedures were headed by literate people, in some hamlets, the prestige and trust that the residents deposited in certain members of the community freed them from the condition of "managing letters." This seems to have happened in the case of Gladis' family. Her grandfather, who was "illiterate" in formal terms, is mentioned[12] as the power behind the opening of the Bañado School. With the help of literate neighbors and the teacher of a nearby hamlet, they drew up documents explaining the need for the school to open, and through them went to the General Department of Primary Schools and the Provincial Education Council. As the local political leader, it was he who then also brought to the attention of the political boss of the Department and of the local Senator a census showing the existence of school-aged children. The lack of formal knowledge of reading and writing did not prevent Don Julian from participating in literacy practices and thereby being the representative of his community in all these proceedings, and then becoming president of the school committee and the cooperative.

The fact of "being comfortable" with bureaucratic requirements, for which this man was recognized, had nothing to do with the formal, schooled definition of the ability to write and read but with the ability to handle negotiations related to these, including recognition and use of written texts where needed. The social and political capital he was able to deploy made the written documents more efficient for the community and these kinds of capital were resources which other members of the hamlet did not possess, even if they were "officially" literate. Don Julian was also the "guardian" of the documents that the community considered important to protect, and later his family kept some of those papers.

The documents we have consulted in the Archives of the Directorate of Primary Schools and the stories of the families also mention other negotiations taken to the Ministry of Education from the peasant communities studied. These called for the removal or punishment of teachers in times of conflict or when for various reasons they did not meet the expectations of the parents. The "repeated absences of teachers," "mistreatment" of children or families, the refusal to "loan" the school for holding parties, and other events gave rise to written complaints and requests for solutions presented by the parents. Other examples of relations with the state mediated by writing, to access goods and services and to assert their rights, are requests or complaints about the conditions of roads, cleaning of canals, or provision of health services.

As Bourdieu (1990) puts it, a good becomes capital to the extent that it presents as "rare" and worth being sought after and valued as a resource in a given social formation. Being scarce, it produces interest by its accumulation. Around it, a certain division of labor is also established between those who produce it, consume it, distribute it, and legitimate it.

The ability to read and write, familiarity with the documents and petitions or "know-how with papers" were, in the peasant families studied, scarce goods and became significant cultural capital that helped maintain or improve the social position of the family and their position in the network of local power. And this was especially true as communications with the outside progressed, services and products arrived, and migration strategies increased.

At the stage where farm self-sufficiency predominated and communications with other hamlets, towns, or cities were more scarce, there seem not to have been many practices of written records of production and labor activities "because for the work that there was, this was not needed." Literacy began to be more important for the families when relationships with the labor market, consumption, and products increased. When, as Rolo mentioned in a quote above, the contractors arrived for seasonal work in the Pampa area, the men who were joining the contingent of harvesters had to sign papers tying them to the "bosses" for several months a year. Something similar occurred when traveling salesmen arrived, bringing "merchandise" that could be exchanged for local production (especially baby goats and corn), or when the men of the community began to leave for the towns and cities to sell the leather or meat of their animals.

Don Juan, the grandfather of one of the study families, explained this when he commented that commercial transactions demanded certain knowledge that had to be managed to ease the operation and avoid "being cheated" by the traders who supplied them or to whom they delivered their products. Managing reading and writing and summing would become a resource and a sign of distinction that allowed him to position himself better in his business.

> When you're in a business like this [the sale of goats], it's worth knowing. It's not the same to arrive [at a wholesaler or dealer of the city of Dean Funes or Córdoba] and be told, let's see, how many animals, how many kilos, how much leather? And to check and control, as when you didn't know and they put one over you … So it was always an advantage, it looked better, because you came from the country, and they said, what do they know? But this way they respected you …
>
> *(Don Juan)*[13]

This same man described the process through which he learned to read and write. Juan did not attend school, but "learned of letters and numbers" while working with the owner of a store in his hamlet and during migration experiences in the south of Santa Fe Province.

> The first one who taught me was Don Toledo when I helped him in the store. I used to spend lots of time there, working, and he taught me little by little … I got better at it over there near Venado Tuerto. Where we used to go, there were some guys working who, in the evening, when we were not in the field, used to come to the sheds and made us practice … They talked about politics … they would tell me they were socialists … I already knew a bit and was enthusiastic and kept practicing. The boys told me I had a head, and not to give up, and I liked it, and it's a question of practice till you don't forget anymore.[14]

The experiences of migration for seasonal work in the harvests or the longer and stable migrations to the cities were thus also spaces for acquiring knowledge about reading and writing and possibilities of participation in literate events.

In the case of the men, another space they mention is the "colimba."[15] The young people who arrived at compulsory military service without having gone through school or having had such a short time of schooling that they had not learned to "deal with the letters" had the opportunity to access this instruction.

For girls raised in the countryside, the first migration experiences were generally to cities to join the informal labor market as domestic servants. In our interviews, this job figured as a turning point in their lives, as it was the first time they left their families and had their own income; but they also remember it as giving them new knowledge and opportunities. One that they mention has to do with contact with the "literate world." The need to follow shopping lists, take messages, be able to read notes with instructions from their bosses, get around the city, have daily access to magazines and newspapers, and to write letters to their families was the impulse for some of our respondents behind the need to acquire or reinforce knowledge of letters.

Finally, in relation to the latter, in all the study households, the letters sent by family members who had migrated played a significant role in mobilizing the relatives. When there was no member who could read and write, they had to find help to enable communication. But, in any case, this correspondence was a precious commodity that the families kept, to some of which I had access in my research.

Final Reflections

I noted at the beginning of this chapter that the NLS offers a theoretical framework for understanding how literacy practices are socially constructed, and that we found points in common with the perspective of Pierre Bourdieu (Grenfell et al., forthcoming). Both of these offer us conceptual tools that help to understand how written culture becomes inserted in the lives of people and how the structural constraints and power relations connect with their needs, interests, and experiences.

In the preceding pages I tried to show how literacy practices can be analyzed as part of the social reproduction strategies that peasant households implement to maintain or better their positions in the system of class relations. This involves considering various analytical, objective, and symbolic dimensions within a historical perspective. It is a matter of reconstructing the history of these families, which in reality always refers to a collective history, through which the social conditions and opportunities are constituted for the use and appropriation of literacy.

Access to reading and writing by peasant families in northern Córdoba was linked to the school. However, it was not the result only of the presence of the State, of the establishment of the educational system, and the imposition of compulsory schooling, but was related to other structural processes, including changes in agrarian structure and in the labor market. As time passed, the objective conditions changed (their positions in the economic, social, cultural fields), but also, along with the transformation of dispositions acquired throughout the histories of migration and linkages with the market and the State, new interests were steadily configured that enabled the incorporation of written culture as a capital, an efficient resource that the children have to possess and could then exchange in the new market place.

In these pages it was possible to present only some of the spaces of literacy that were configured in the life of the peasant families of Tulumba. They show, however, a richness and complexity that is often overlooked in educational environments that repeatedly qualify rural spaces as "illiterate." What is also interesting is to note how the interviewees themselves appeared to ignore this wealth and recognized only the school as a literate space.[16] This invisibility is also part, as both the NLS and Bourdieu indicated, of the networks of power relations that must be analyzed in the different empirical cases.

Notes

NOTE: The author of this chapter has used pseudonyms throughout to protect the identities of students and instructors

1 See, among many others, Street (1984), Street (1993), Barton and Hamilton (1998). The Spanish version of the central NLS texts can be seen in Zavala, Niño Murcia, and Ames (2004).
2 We follow the definition of access by Kalman (2010, 11). She distinguishes "the material aspects from the social ones through the notions of access and availability. While availability refers to the material conditions necessary for literacy, access to literacy goes beyond the presence of print (or screens); learning to read and write requires the opportunity to interact with other readers and writers, to talk about written text, to insert its use in multiple situations and contexts, to experience its use and tensions, and to use written language for ones own purposes."
3 See Castanheira, this volume.
4 Source: Anuario Estadístico de la Provincia de Córdoba, Gobierno de la Provincia de Córdoba, 1930. [Statistical yearbook of the Cordova Province, Government of the Córdoba Province, 1930].
5 Source: Tina interview 2 (May 8, 1998).
6 Source: Rolo interview 2 (August 8, 1997).
7 Source: Tina interview 2 (May 8, 1998).
8 Source: Rosa interview 3 (September 6, 1998).
9 Source: Mercedes interview 4 (June 14, 1997).
10 Source: Gladis interview 5 (March 14, 1998).
11 Barton and Hamilton (1998) describe literacy events as observable activities in which reading and/or writing develop. These activities are always embedded in social contexts and emerge from literacy practices that constitute cultural forms for using the written language. Street (1993) adds that, unlike events, practices are not directly observable since they include a series of values, attitudes, feelings, and social relationships. See also Nabi, Rogers, and Street (2009).
12 The reference to Don Julian's activities in relation to the opening and management of the school is made not only by family members during the interviews but it is present in official documents, Records of General Inspection of School Buildings, and Inspection Records of Tulumba department. Years 1931–1944. Archives of the Directorate of Initial and Primary Level Education of the Province of Córdoba (DGEP).
13 Source: Juan interview 1 (May 8, 1998)
14 Source: Juan interview 2 (August 14, 1998)
15 Compulsory military service called young men to military training and to provide various services in the Armed Forces. In the beginning, people were recruited between 20 and 21 years of age, and it lasted 18 to 24 months. From the 1970s, 18-year-old men were incorporated. It was popularly known as the "colimba," a word that alludes to three frequent conscript activities: running (*corre*), cleaning (*limpia*), and sweeping (*barre*).
16 Nabi, Rogers, and Street's subjects also referred to themselves as "illititerate," even though they could be shown evidence of their actual uses of everyday literacies. See Nabi et al. (2009).

References

Barton, D., and Hamilton, M. (1998). *Local Literacies: Reading and Writing in One Community*. London and New York: Routledge.

Bourdieu, P. (1988a). *Cosas dichas* [*Things Said*]. Buenos Aires: Gedisa.

Bourdieu, P. (1988b). *La Distinción: Criterios y bases sociales del gusto* [*Distinction: Criteria and Social Bases of Taste*]. Madrid: Taurus.

Bourdieu, P. (1990). Algunas propiedades de los campos [Some fields' properties]. In *Sociología y Cultura* [*Sociology and Culture*]. México: Grijalbo, 135–41.

Bourdieu, P. (1991). *El sentido práctico* [*The Practical Sense*]. Madrid: Taurus.

Cragnolino, E. (2009). Condiciones sociales para la apropiación de la cultura escrita en familias campesinas [Social conditions for the appropriation of written culture in rural families]. In J. Kalman and B. Street (eds.), *Lectura, escritura y matemáticas como prácticas sociales: Diálogos con América Latina* [*Reading, Writing and Mathematics as Social Practices: Dialogues with Latin America*]. Mexico City: Siglo XXI Editores and CREFAL.

Grenfell, M., Pahl, K., Rowsell, J., and Street, B. (forthcoming). *Language, Ethnography and Education: Bridging New Literacy Studies and Bourdieu*. New York: Routledge.

Kalman, J. (2010). Literacy Learning in Community Contexts: Plazas, Computers, and Sewing Rooms—Latin American Perspectives. Paper presented at the 2010 Annual Meeting of the American Educational Research Association, Denver, Colorado, April 30–May 5.

Nabi, R., Rogers, A., and Street, B. (2009). *Hidden Literacies: Ethnographic Case Studies of Literacy and Numeracy from Pakistan*. Bury St. Edmunds: Uppingham Press.

Street, B. (1984). *Literacy in theory and practice*. Cambridge: Cambridge University Press.

Street, B. (1993) (ed.). *Cross-Cultural Approaches to Literacy*. Cambridge: Cambridge University Press.

Zavala, V., Niño Murcia, M., and Ames, P. (eds.) (2004). *Escritura y Sociedad: Nuevas perspectivas teóricas y etnográficas* [*Writing and Society: New Theoretical and Ethnographic Perspectives*]. Lima: Red para el Desarrollo de las Ciencias Sociales en el Perú [Network for the Development of Social Sciences in Peru].

4

A MULTIMODAL APPROACH TO THE UNDERSTANDING OF STUDENTS' COLLABORATIVE WRITING OF DIGITAL TEXTS

Enna Carvajal

Ms. Garcia, the chemistry teacher in a senior high school in Mexico City, talks with a group of three students about an assignment she has given them. She is asking them to review the written instructions she has handed out to make sure that the girls understand what she has asked them to do. In the following exchange, she is discussing with one of the students how to elaborate a poster for a school science fair:

Ms. Garcia: Will all the posters be the same size, or …?

Diana: Yes, put it there … look … maximum size 75 cm x 1.20 cm [reading from the instruction booklet].

Ms. Garcia: Yes, because then they don't fit in the spaces. Remember that a lot of text doesn't look good.

Diana: Yes, we have to make one tomorrow.

Ms. Garcia: Then learn from this one too, to do it better. What I saw last year was there were very pretty photos, of solar energy for example, but their poster didn't tell you anything at all, unless they explained. It has to be self-explanatory, though the presenter complements it; if there's no information, it's an advertisement, not a poster.

Diana: In fact, they tell you: don't include too much information; don't use graphs or photographs that have no explanation [regarding the instruction booklet for the poster].[1]

The above dialogue foregrounds the role played by the instructions given to students in senior high schools during their writing processes when trying to complete their assignments (Méndez, 2004; Nelson, 1990). It is an invitation to look more closely and examine how students ponder teacher guidelines in their efforts to successfully produce what is asked of them. In the context of school, students learn to redefine their notions of what is important for a particular course or a specific assignment, even when the teacher is not explicit in this regard.

In the case in question, the teacher defines the criteria for acceptable academic work and explains how traditional modes of evaluation place stress on written and printed content as the valid means of representation of academic contents, notwithstanding the wide range of equally valid and powerful means offered by the variety of modes—visual, aural, animated—evident in the use of computing tools (Jewitt, 2009). To understand how the different options for creating meaning in student interactions and writing operate, in this chapter I will focus on the different symbolic means of representation—i.e. writing, image, and other resources that each student uses for representing and communicating meaning.

I will argue that the definition of a learning activity shapes students' uses of available resources, different modes of representation, and technology as a mediating tool. When the main activity students are involved in is searching for information, the use of technology as a mediating tool is sometimes reduced to a purely instrumental practice for retrieving; copying information from one text and pasting it in another (see Guerrero, this volume). When students are faced with an activity involving the elaboration of a specific genre, such as a poster, they must manipulate and transform the information. In this case, their use of the computer as a mediating tool extends to other properties: it is used to reorganize, synthesize, reformulate, move from one mode of representation to another, changing how the original information was originally presented. This change generates a deliberate manipulation of the content that is reflected not only in its reorganization, but the decision of how to embody and display it.

This chapter stems from a broader study that set out to describe and analyze, from a socio-cultural and interpretative viewpoint, the way a group of senior high school students in Mexico constructs meaning via computer-mediated collaboration and expresses their understandings of the academic products requested by their teacher as part of a science assignment. The research focused on exploring the interactions and negotiations that took place among them, the way in which they reached agreements, how they used the resources at hand, how they conceived of the assigned activity, and how all these factors affected the drafting of the participating students' academic texts.[2]

A central premise guiding the analysis of the students' participation in these activities and their writing process comes from theory developed as part of the ethnography of communication agenda: it is assumed that the high school girls working together shared knowledge about the local tenets guiding interaction and interpretation—the rules of engagement in academic activities—and that their understanding of these rules shaped their choices and defined their selection of different forms of expression consistent with the social expectations of their particular context (Gumperz, 1982; Hymes, 1972). New Literacy Studies (NLS) also provides a theoretical framework focusing on the participants' strategies for understanding, discussing, organizing, and producing their written text and their poster according to the discursive practices of the school community where they belong (Street, 1984).[3] This approach assumes that a complex relationship exists between the students' local productions and broader dominant academic practices (Street, 2005). Both theoretical frames—NLS and the ethnography of communication—emphasize the importance of context and participation for understanding language and literacy practices and the development and exchange of situated meanings.

The assignment's design also defined which options the students had available to them as well as the expected results. It mediates the teacher's expectations of what she assumes her students will learn and the students' perceptions of what they are being told to do. Studies of the processes involved in the production of texts such as the one here proposed examine both the instructions and perceptions related to the assignment, the expectations regarding the end product, and the use of available resources (Goodfellow, 2005; Jones and Lea, 2008; Levy and Kimber, 2009). They identify the role played by academic writing practices in digital environments within a complex context of multimodal representations.

In this specific case, the students use and combine different modes of representation to cope with their assignment, which involved the design and elaboration of a poster related to the topic of soil pollution. The final destination of their work was the science fair and therefore was to be conceived as an exhibit. In the course of their interaction, I observed how the students used language in order to make their contributions to the collective activity—contributions that were molded and fit into the texts they constructed—and reflected their shared understanding of the nature of the activity, the academic content, and its definition in terms of the particular context. I understand this context to be a complex construct in which school directives, institutional conditions, and fleeting interactions among the participants converge, requiring a theoretical vision that contemplates the social, discursive, and phenomenological dimensions of language (Kalman, 2010).

To address these issues, I assume that the literacy practices can only be analyzed in terms of their relationships to a specific context (Duranti and Goodwin, 1992; Gumperz, 1982; Street, 1984). For that reason, I center my attention on the use of language in the situation in which it is produced, taking stock of the complex situations in which communicative events occur via written texts, and their interactive, cultural, interpretative, and ideological dimensions (Heath and Street, 2008). In turn, the students' local discussions and meaning making can be linked to broader contexts of academic conventions and practices related to interpreting and producing the contents of science (Brandt and Clinton, 2002).

Given the use of the different resources available to the participating students—e.g. those offered by computers and the programs they use to create their texts and posters—the study of the semiotic design of multimodal resources that provide conceptual and analytical tools for analyzing representations becomes relevant for understanding how students make meaning in the context of their assignment (Bezemer and Kress, 2008).

Multimodal theory is a conceptual tool that allows the analyst to show how meaning makers use and combine resources and how different factors give rise to specific forms of textual cohesion and coherence—e.g. the resources available, the influence of dominant textual genres, and the types of expectations that intervene in the relationship with the audience, among others—are interwoven (Bezemer and Kress, 2008). What follows is a brief presentation of some of the central aspects of a multimodal analysis, illustrated by specific examples from this study (see Lorenzatti, this volume).

Multimodal Theory as an Analytical Frame

The assignment the group of students received was a complex one. It consisted of a research project on a scientific topic (soil contamination) and the elaboration of a series of products, including both a written text and then a poster to be collectively made by the participating students. In this chapter, I center my interest on the creation of the poster, because the observations of this particular activity made apparent the students' development of its design, first as a report, then as a *PowerPoint*® presentation, and finally as a poster for display. This process revealed the software's features for quickly offering different modes for reconstructing meanings in the presentation screens that the students had already expressed in the previously written research report. In other words, reformulation and manipulation of the academic contents occurred as students made changes in its format, suggesting that the participants take various design factors into account in compliance with the discourse rules they are expected to fulfill. In the course of their project, the students exploited the resources available to them through digital technology (the malleability of electronic texts, whereby paragraphs can be added, formats can be changed, and easily edited; tools for tracking changes; and the inclusion of comments), interacted, and jointly constructed their poster.

The multimodal nature of digital tools permits the incorporation of image, sound, and animation, as well as instantaneous links with other texts. All of these options break with the more rigid, traditional, linear, sequential layout of printed text, which does not allow any of the above actions to be carried out at the speed and with the ease that electronic media allows.

Students brought formatting resources such as font type, color, size, etc. into play alongside the premeditated manipulation of copied and pasted contents in different fonts (i.e. from web pages). The first form their project took was a lengthy document, where they collected and listed different fragments:

Diana: Okay, no . . . I will copy this [referring to the first paragraph of a document on soil contamination which she is reading].

Alma: But you have to sum it up because if we put that ...

Elvira: Yes, we should only put in the basics.

Diana: [She copies and pastes the information into the presentation slide and begins to read it] *From the soil* ... Inappropriate is far fetched [deleting the word] ... *management of solids and liquids* ...

Elvira: But isn't this just a continuation of the first idea?

Alma: Yes.

Elvira: Then with the first idea we have enough.

Diana: It's too long to be an idea.

Elvira: [Reading the idea] *Life cycle of the soil obeys the rules of an ecosystem composed of an inorganic mineral substance that is ... Just as plants are able to produce organic matter through photosynthesis and that ...*

Diana: I mean, *the life cycle of the soil obeys the rules of an ecosystem*, period.

We can assume from the previous exchange that the written document the students produced as part of their research report was subjected to revisions, summarizing, paraphrasing, or redrafting, in order to satisfy the *PowerPoint®* design requirements for the poster. This includes making decisions about space, contemplating how much time will be available for the oral presentation of the poster, and its evaluation by the teacher. In the course of this poster-production process, one can observe the features for designing that make the construction of the text genre possible and the students' perceptions of the teacher's guidelines for the activity and available resources.

Every usage of electronic media for the purpose of reading, writing, and meaning construction invites us to reflect on the nature of multimodality: meanings are created, distributed, received, and interpreted via different modes of representation and communication, even in tasks that do not imply the use of electronic media. Multimodal analysis provides us with tools for analyzing and describing the wide range of resources that are available for the purpose of building and organizing meaning (Jewitt, 2009), as in this case, where the content of a written research report needs to be rearranged to fit requirements and characteristics of a poster. Furthermore, students are exposed to a certain tension when having to deal with a set of not-so-explicit ways of evaluating the academic knowledge and its forms of representation that they are expected to produce in their written documents—e.g. the use of bibliographic sources, adherence to academic conventions, and the inclusion of expressions and repertoires that are deemed acceptable by the academic community (Blommaert, 2008; Johns, 1997; Lea and Street, 2000). The influence of these dominant notions and genres on the reproduction of traditional practices may even inhibit the use of the multimodal resources that technical tools make available to us (Jewitt, 2009; Kress and Bezemer, 2009; Street, Pahl and Rowsell, 2009).

From the multimodal point of view, a digital text contains meanings that are determined by a range of modes: here, "mode" is a socially and culturally molded resource that enables meaning to be constructed. Some examples of modes are images, written discourse, video, and diagrams, in each of which one can identify resources. For example, a written text possesses syntactic, grammatical, and graphic resources such as letter type, size, color, and punctuation, all of which constitute modal resources (Bezemer and Kress, 2008), while intermodal interaction is part of the production of meaning, which, indeed, depends on the operational rules that pertain to a specific social context and the motivations and interests of the person(s) who produce(s) it (Kress, 2009). Insofar as it relates to the aims of this chapter, multimodal analysis provides ways to study reading and writing practices using the computer, which are characterized by continuous translations between meanings that are constructed in diverse modes (Kress, 2009). The movement from one mode to the next implies representing meaning in one semiotic register and then re-expressing it in a second one (Duval, 1999).

While intermodal changes augment representational power, they also demand the consideration of features or resources that do not have to be considered in other modes—e.g. changing from a lengthy written document to a *PowerPoint®* presentation, which may involve images or video. Indeed, according to Matthewman and Triggs (2004), manipulation and negotiation regarding the use of modal resources may aid some aspects of the joint construction or shared activity. These authors note that students spontaneously begin exploiting the resources that a given tool has to offer to the extent that they see the benefits in taking advantage of the potential inherent to the various resources and modes of representation.

It can be said that the affordances of a technological resource are being fully exploited when students use them successfully according to their aims, no matter whether these be communicative or pedagogical, recreational or academic. This situation largely depends on the context of use. The redefinition of the contextual conditions of a teaching situation may thus enhance students' perceptions regarding the potential uses of certain resources and help to achieve the pedagogical aims pursued in the said teaching situation. Contextual conditions are redefined through the ways activities are set up and evaluated, what aspects of student work are evaluated, and the types of interaction and participation teachers promote with and among their students.

The interests of the students as designers of certain texts and their understanding of the makeup of the target audience determine the potential use of resources, and the alignment and coherence among them. Even when this point of view has been increasingly adopted in research projects (Jewitt, 2009), in present-day teaching curricula, linguistic representation in the form of reports or essays is privileged in formal education and evaluation, especially in Latin American schools. Though other multimodal means of expression may be contemplated for teaching purposes, such modes are underrepresented when it comes to work developed by students and its evaluation, and this situation may influence students' choices when deciding on what modes of representation to use and deploy. The assumption seems to be that writing and certain text genres are more in compliance with institutional requirements than other modes of representation, which they believe may not be equally acceptable.

The Students' Activity in Context

In the course of creating their poster, the students in this study used various resources available to them (Blommaert, 2008); their selection of different options enabled them to give meaning to their joint activity, and to the work they produced. Most of the time, these young people interacted vis-à-vis word-processed texts displayed on a computer screen. Starting with a text composed of fragments that they copied and pasted from various web pages, they amended and included this material using word-processing resources—i.e. selecting, copying, and changing files, pasting paragraphs, modifying colors, underlining sections, inserting comments, and formatting the text, among others. Throughout their work sessions, the participants discussed different aspects of their text and manipulated fragments of information within the new textual context (Blommaert, 2008), writing in new passages based on the oral exchanges that occurred when revising together.

In order to create the poster to be displayed at the science fair, the participants first used the computer to put together a *PowerPoint®* presentation that they would then print and glue directly on to the poster board. Following the instructions on how to produce a poster with the correct dimensions, the participants used two sessions of two hours each to design the poster; as part of planning and organizing it, they created a working document in the form of a table on the word processor showing how many slides they should create and how the content should be distributed across each one.

It should be stressed that one text genre was being transferred to another—i.e. a research report written in continuous discourse typical of an academic document was being transformed into slides to be used on a poster meant for public display, and also an electronic means of representation was being turned into a printed poster. In this case, the transfer implies manipulating resources such as color and images and taking the available space into account. The creation of the poster involves processes such as the summarizing and paraphrasing of content, and the students' understanding of the pertinent academic contents is reflected in their decisions as to how many slides they need to cover each part of the poster's contents. Among the elements that the teacher had asked the students to include, the scientific–technical ones would appear to merit a privileged place in the presentation (three slides vs. one slide for each of the other elements).

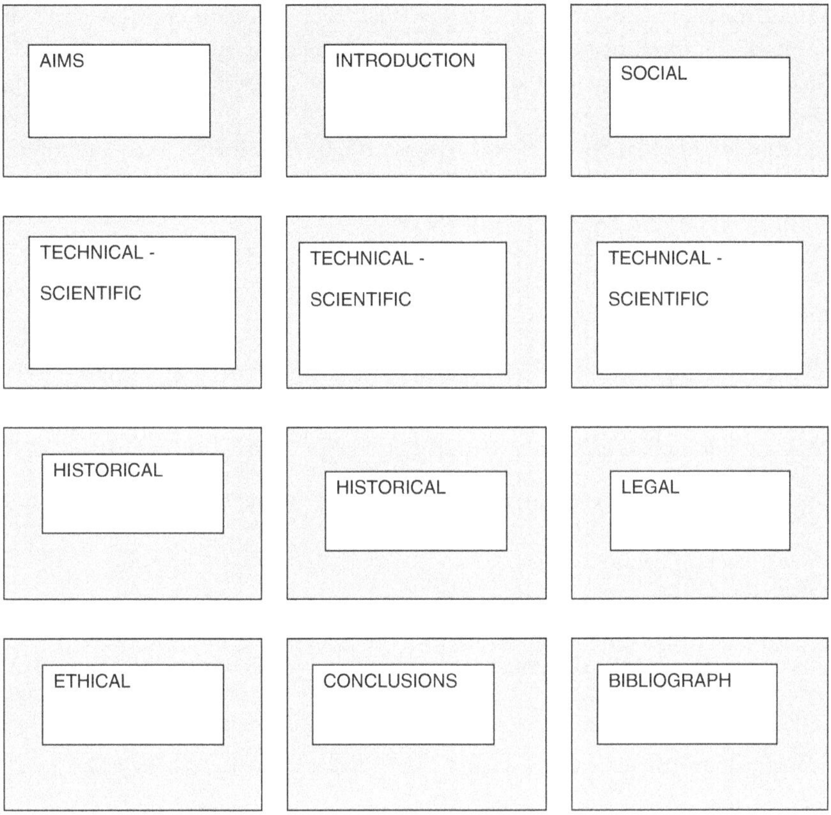

FIG. 4.1 Table created by the students showing the design of the poster for the science fair.

The participants' prior experiences with regard to research presentations can be inferred from their linkage between the number of slides and the planned length of the presentation, and this bears witness to the ease with which they are able to tackle this specific genre (poster/presentation) within an academic context.

The work done by the students to design the *PowerPoint®* presentation is a concrete example of how they exploited the available repertoire of resources: in this case, the program in question, traditionally used for projecting electronic presentations, serves as a useful tool for creating printable objects to be publically displayed in the form of a poster, making use of its design resources (the inclusion of images, the handling of color and the fonts), and its different product options (digitally animated screens, outlines or printed illustrated texts). The participants negotiated and reached agreements about some aspects of their poster for the science fair, and, in the next exchange of opinions, the discussion focused on both the background color of the slides and the text that each screen was to contain.

During the activity, the students collaborated as they moved through the different stages of their project; their collaboration was facilitated by their shared criteria regarding the requirements and the meaning of the assignment. In the following example, elements pertaining to form are placed at the service of content to be communicated (Matthewman and Triggs, 2004). In terms of the multimodal aspects that students work with here, letter type, background color, and the color of the letters related to the background can all be identified as a modal resource that favors meaning making (Kress and Bezemer, 2009, 7).[4]

Alma: I like strong fonts.

Diana: Exactly, and clear letters. I feel that they stand out more; they're more attractive.

Alma: I don't like black letters in projects.

Diana: A color … I mean it can't be red or orange, because they're not serious.

Elvira: Yeah!

Diana: I like this one [choosing a dark shade of blue and changing the background of the slide they are working on, which still has no content].

Alma: Yes, OK.

Diana: Now it's any type of letter … but I think white looks very good.

Elvira: Yes … with white and wide [referring to the letters].

Diana: I feel that the letters should be serious and fat.

Elvira: **Verdana and bold** [still referring to the type and color of the letters].

Diana: OK, you can see it because you can see it!

Elvira: And it looks good.

Some relevant aspects of this exchange have to do with the degree of contrast between the background color and the color of the letters. The intention is to draw the reader's attention to the text by using color and make it easier to read, within the limits imposed by the students' understanding of the seriousness of the project—i.e. colors considered inappropriate for an academic project (red and orange) are eliminated as possible choices. Finally, the participants' summary of what they have agreed on stands out: in concrete terms, the letter type chosen has to be "serious and fat," meaning "**Verdana and bold.**" Through this choice of font (verdana), its color (white) on a blue background, and its particular style (bold), the students try to represent what they consider an academic text, establish a connection with the audience, other members of their academic community,

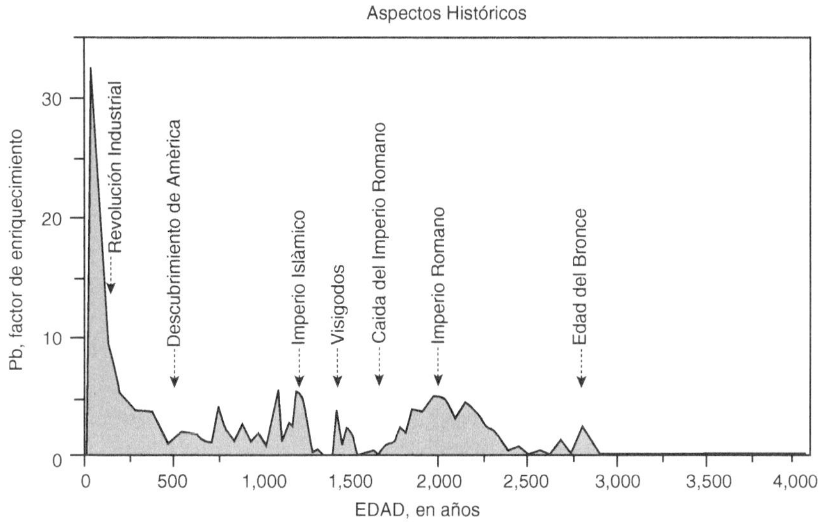

FIG. 4.2 Slide 3 of the poster: Graph of lead content in soil over time.

and fulfill the teacher's—who will eventually evaluate the project—directions (Jewitt, 2009).

In the conversation that opens this chapter, the teacher spoke with one of the students about how to design the poster and what specified her expectations. There the teacher defined what she considered to be a good use of space, the amount of text each screen should include, and the use of images and graphs. With regard to this requirement, a conflict arises among the students when they discuss how to include a graph downloaded from an internet page that represents some of the historical aspects of pollution in the poster.

In the next fragment of student talk, the participants discuss how to write an explanation for the graph:

Elvira: OK. What are we going to do with the information?

Alma: Because we can explain it, but it's understandable just as it is, without anything else.

Diana: And precisely what she said was "You can't include images if they don't include an explanation" [referring to the teacher's instructions on how to design the poster].

Elvira: And is there no way we could make the graph look good [meaning could the graph be made smaller without getting fuzzier]?

Diana: With an image editor, maybe, but … we don't have one.

The participants finally solved the problem of not having enough space to include an explanation of the graph by adding another slide where they put the information about the historic aspects of pollution. For the students, there was a tension between the idea (apparently shared by them) that the image was self-explanatory and the teacher's instruction to include an explanation for every image used in the presentation. Furthermore, they had to deal with the amount of text that can be included on any given screen. The reader will recall that in the opening lines the teacher had stated "Remember that a lot of text doesn't look good."

In a solution that is a compromise between the two options (footnoting the image or using the image without explanation), instead of creating the specific footnote, the students hoped the graph would be understood in the context of a further slide that describes the historical aspects of pollution.

The students' purpose for including this image was to show variations in levels of soil contamination over time in accordance with variations in human activity. They are more confident in the explanatory capacity of the image than in that of the text—i.e. the visual mode is deemed to be a better means of getting a given message across than the textual one (Jewitt, 2009). According to Jewitt, students generally favor one mode over another when "reading" multimodal texts online, and most of them prefer images. In this case, this same preference can be observed in the production of their own texts.

Moreover, though the web page from which the graph is taken contains a long, detailed, technical explanation, the students focus on the scope of their assignment and their own understanding of what they are doing, so they decide to use only the image, separating it from its original explanation in order to convey a more general idea in line with their project. In their own multimodal design, they transform or reformulate the meaning of the image (Jewitt, 2009); they reconstruct it according to both the content requirements of the assignment and their perception of the image's capacity to depict the content in question.

Conclusions

Considerations of space have limited this chapter to the presentation of just two examples of the important work students do when manipulating multimodal aspects of the representation of meaning during a collaborative activity involving the production of a poster using the computer. Nonetheless, I believe that these examples provide evidence of the weight assigned to information-bearing content when doing a school research project: despite the plethora of resources such as images, audio tracks, and hyperlinks in the documents that the participants consulted via the Internet and selected, copied, and pasted for their project, the linear nature and sequential practices that they employ and the final decisions they make concerning what to include in each screen and how to include and reformulate their selections are surprising. Even though they were creating abbreviated texts that were to be part of a poster, what they choose to do in the end is closely linked to standard notions of printed academic texts, where written discourse is the dominant feature. They included only two images in the final version of their poster.

In the first example, one is struck by the students' ability to distinguish between different designs' features as modes or vehicles of meaning, capable of representing a particular message. Albeit within the framework of the formal requirements that the students must satisfy, formatting resources such as letter type, color, size, etc. are brought into play, along with the deliberate manipulation of content for the purpose of complying with design requirements that involve decisions about space, presentation length, and evaluation.

The second example, particularly, shows how the graphic component is limited due to the restrictions imposed by the teacher, the definition of the activity, and the students' conclusions based on prior experience when working on similar projects. The participants are faced with a difficult decision, since there is tension between their need to adhere to the guidelines and the value they assign to graphic resources.

The pressure exerted by the rules of interaction determines the products, which are laden with information and favor a genre that is linked to conventional written text assignments. Though the use of the computer for researching, reading, and constructing meaning (Jewitt, 2009; Matthewman, Blight and Davis, 2004) implies pondering multimodal representations, the participants' products nevertheless pertain to a linear, sequential form of content based on the requirements governing written academic assignments. There is something artificial and contradictory in the participants' immersion in the web to search for information that will bring them into contact with the multimodal nature of digital texts that they are later obliged to eschew when performing the academic work assigned to them.

If, as Gee asserts (2003), students' sophisticated manipulation of multimodal representations in everyday contexts endows them with a strong foundation on which to build proficiency in interpretative activities and become involved in the multimodal conveyance of meaning, then our common school assignments such as the one analyzed here are failing to give students the chance to put what they have learned into practice.

The use of other modes of representation, like image or video, provides us with the opportunity to express meaning through them, both singly and in combination: in the present case, the decisions taken with regard to the inclusion of images in the project reveal conventional attitudes, notwithstanding the students' positive appraisal of the ability of the graph that is included in the poster to transmit meaning about the historical aspects of pollution.

To consider that some students who enter formal learning contexts are already familiar with digital texts and can move freely among them has also implications for the evaluation of multi-format materials and their communicative aspects. It also constitutes a challenge for teachers and educational systems because it is a new situation that conventional didactic designs fail to take into account.

While the computer that the students use to do their assignment enables them to extend and enhance their text-building and poster-designing activities by providing them with tools that give almost instant multimodal information on the set theme, programs for manipulating information and representing it in different forms, and resources for sharing their products and exchanging information with others, their opportunity to exploit these options depends not only on the availability of these resources and on the extent to which they can adapt or use them meaningfully but also on the legitimization of multimodal expression in the school context and for academic purposes. It does not suffice for teachers to design assignments with set resources and suggest their use: they have to follow up on this as well by ensuring that students use them, and take stock of unforeseen, spontaneous uses.

My analytical approach to the ways in which students make meaning from the information that they get from the Web, based on the aims of their assignment and other contextual delimitations, distances this study from those that place more analytical emphasis on outcomes than they do on exploring students' processes or tracking the ways they transform the information they gather in the course of creating those outcomes. In this regard, an analysis of how information is organized and worked with reveals the selecting, copying, pasting, summarizing, paraphrasing, and resource-selection procedures, etc., and makes these phenomena visible for teachers and researchers alike. The findings from this line of research can potentially allow teachers and program designers to keep the use of technology, key to this process, in the forefront, and provide a basis for developing concrete guidelines for students. At the same time, it provides important insights regarding the different kinds of texts that students might create to communicate academic content. The ability to transform and improve educational practices lies not in the tools themselves, but in the uses to which participants put them when tackling content and engaging in learning activities.

Notes

NOTE: The author of this chapter has used pseudonyms throughout to protect the identities of students and instructors.

1 In all cases, fragments of interactions have been translated and all transcription codes have been removed to facilitate reading. The use of basic punctuation marks relates to their conventional intonation use. Ellipses indicate pauses in the dialogue; brackets include clarifications or actions. Italicized indicates reading texts.

2 This contribution is part of the research described, a cross-cultural collaborative project involving six senior high schools in Mexico and Canada, during school years 2001 to 2004.

3 In order to study the students' choices in the process of creating academic products, I collected hard copies and electronic examples of the students' texts, kept field notes, and integrated them with transcriptions and coded versions of tape-recorded interactions. This rich database enabled me to describe and analyze observed events, explain some of the social processes involved, and unpack which kinds of academic

requirements shaped their engagement with the activity and their use of multimodal resources.

4 Different conceptions exist of the concept of mode, depending on who constructed the meaning, and in accordance with the needs and interpretations of a given community that adopts certain conventional practices (Kress and Bezemer, 2009, 7).

References

Bezemer, J., and Kress, G. (2008). Writing in multimodal texts: a social semiotic account of designs for learning. *Written Communication* 25 (2), 166–95, http://wcx.sagepub.com (December 2009).

Blommaert, J. (2008). *Grassroots Literacy: Writing, Identity and Voice in Central Africa.* New York: Routledge.

Brandt, D., and Clinton, K. (2002). The limits of the local: expanding perspectives on literacy as a social practice. *Journal of Literacy Research* 34, 337–56.

Duranti, A., and Goodwin, C. (eds.) (1992). *Rethinking Context: Language as an Interactive Phenomenon.* Cambridge: Cambridge University Press.

Duval, R. (1999). *Semiosis y pensamiento humano: registros semióticos y aprendizajes intelectuales* [*Semiosis and Human Thought: Semiotic Records and Intellectual Learning*]. Cali, Colombia: Universidad del Valle.

Gee, J. (2003). *What Video Games have to Teach us about Learning and Literacy.* New York: MacMillan.

Goodfellow, R. (2005). Academic literacies and e-learning: a critical approach to writing in the online university. *International Journal of Educational Research* 43 (7–8), 481–94.

Gumperz, J. (1982). *Discourse Strategies.* Cambridge: Cambridge University Press.

Heath, S., and Street, B. (2008). *Ethnography: Approaches to Language and Literacy Research.* New York: College Teachers Press.

Hymes, D. (1972). Models of the interaction of language and social life. In J. Gumperz and D. Hymes (eds.), *New Directions in Sociolinguistics.* Malden, MA: Blackwell, 35–71.

Jewitt, C. (2009). Multimodality, "reading," and "writing" for the 21st century. *Studies in the Cultural Politics of Education* 26 (3), 315–31.

Johns, A. (1997). *Text, Role and Context.* New York: Cambridge University Press.

Jones, S., and Lea, M. R. (2008). Digital literacies in the lives of undergraduate students: exploring personal and curricular spheres of practice [electronic version]. *The Electronic Journal of e-Learning* 6, 207–16. Retrieved January 2010 from http://www.ejel.org/Volume-6/v6-i3/JonesdandLea.pdf.

Kalman, J. (2010, April 30–May 5). *Literacy learning in community contexts: plazas, computers and sewing rooms.* Paper presented at the 2010 American Educational Researchers' Association Meeting, Denver, Colorado.

Kress, G. (2009). What is mode? In C. Jewitt (ed.), *The Routledge Handbook of Multimodal Analysis.* New York: Routledge, Taylor & Francis, 54–67.

Kress, G., and Bezemer, J. (2009). Writing in a multimodal world of representation. In R. Beard, D. Myhill, M. Nystrand, and J. Riley (eds.), *The SAGE Handbook of Writing Development.* London: SAGE Publications, 167–81.

Lea, M., and Street, B. (2000). Student writing and staff feedback in higher education: an academic literacies approach. In M. Lea and B. Stierer (eds.), *Student Writing in Higher Education: New Contexts.* Buckingham, U.K.: Open University Press, 32–46.

Levy, M., and Kimber, K. (2009). Developing an approach for comparing students' multimodal text creations: a case study. *Australasian Journal of Educational Technology* 25 (4), 489–508.

Matthewman, S., and Triggs, P. (2004). Obsessive compulsive font disorder: the challenge of supporting pupils' writings with computers. *Computers and Education* 43 (1–2), 125–35.

Matthewman, S., Blight, A., and Davies, C. (2004). What does multimodality mean for English? Creative tensions in teaching new texts and new literacies. *Education, Communication and Information* 4 (1), 153–76.

Méndez, J. M. (2004). *El escrito académico: un tejido de textos múltiples* [*Academic Writing: A Fabric of Multiple Texts*]. Mexico: Universidad Pedagógica Nacional (UPN).

Nelson, J. (1990). This was an easy assignment: examining how students interpret academic writing tasks. *Research in the Teaching of English* 24 (4), 362–96.

Street, B. (1984). *Literacy in Theory and Practice.* Cambridge: Cambridge University Press.

Street, B. (2005). *Literacies Across Educational Contexts: Mediating Learning and Teaching.* Philadelphia: Caslon Publishing.

Street, B., Pahl, K., and Rowsell, J. (2009). Multimodality and New Literacy Studies. In C. Jewitt (ed.), *The Routledge Handbook of Multimodal Analysis.* New York: Routledge, 91–200.

PART II

Literacy and Numeracy as Social Practice: Latin American Perspectives

5

GPS TECHNOLOGY, MAP READING, AND EVERYDAY LOCATION PRACTICES IN A FISHING COMMUNITY[1]

Judy Kalman

The Sian Ka'an Biosphere Reserve[2] (SKBR) in Quintana Roo, Mexico, is part of the Mesoamerican Barrier Reef System that extends from Isla Contoy on the north of the Yucatán Peninsula to the Bay Islands of Honduras (Walker et al., 2004). As you drive down the dirt road that enters the grounds, its narrowest points are only 30 meters wide, setting the vehicle on a narrow stretch of land surrounded by water on both sides.

From 2000–2003, marine scientists from a research center in a nearby city studied benthic habitats[3] and their relation to lobster fishing in Bahía de la Ascención located within the biosphere.[4] They engaged local fishers in data collection activities, setting out on the bay in their boats in order to find predetermined sites where sea bottom samples would be collected. The scientists used geographical positioning systems (GPS) while the fishers steered their boats and followed precise instructions from the scientists. As part of their collaboration, the marine scientists also generated and distributed satellite-based maps of the local fishing grounds. The goal of this chapter is to characterize the fishers' incorporation of GPS technology and the satellite-generated maps into their local practices and to understand the local history of GPS use among traditional lobster fishers (Pahl and Rowsell, 2010). Practice is understood to comprise of actions involving technologies, skills, and social knowledge (Scribner and Cole, 1981) as well as what people think and believe about their actions (Street, 1995).

From their encounter with the marine scientists, some of the local fishers began to learn to use the GPS devices, and the researchers became an important resource for them. They played the role of more expert others who mediated the technology use for the fishers. Once the townspeople began to use their devices, they went to each other for advice. When the scientists developed and distributed the maps of the fishing grounds, they also showed the fishers how to interpret them.

Literacy Studies and Digital Technologies

This chapter is theoretically situated within the New Literacy Studies paradigm, which posits that reading and writing are social practices, bound ideologically to institutional contexts, historical processes, and power relations that reach beyond the immediacy of situated reading and writing (Blommaert, 2005; Collins and Blot, 2003; Street, 1984, 1993, 1995). From this point of view, literacy practices involve both the observable actions related to reading and writing as well as their interpretation (Barton and Hamilton, 1998). Access to literacy (Kalman; 2001, 2004) is seen as the appropriation of discourses, the understanding of meanings, the construction of relevant forms of participation within given context as well as learning the intricacies of written conventions, and refers to the contexts, purposes, and actors involved in literacy learning. Associated with literacy are many cultural artifacts that are valued in a specific context, evoked in oral and written language, carried and used in meaning-making activities and part of people's identities and experiences (Pahl and Rowsell, 2010).

Warschauer (2002) has written a thought-provoking analysis of the notion of access, by relating this term to information and communication technologies (ICT) and literacy. He notes that the idea of access is often limited to the possession of technological devices, basically the ownership of a computer. He argues for a more complex notion, stressing the importance of social contexts of practice and the shifts in the use of technology, depending on historical, political, and sociocultural conditions. While he recognizes the presence of material artifacts is necessary, he argues that, in the case of digital literacies, the presence of devices alone does not lead to technological competence, and stresses the importance of equity for all students: equal opportunity can be offered only by constructing needed contextual supports for learning. In particular, digital artifacts become powerful mediating tools to the extent that people learn to use them, and generally learning takes place through interactions with others, around and through technology.

In my own work, I have distinguished access from availability, using the notion of availability to signal the material aspects of literacy (the physical presence of print materials; infrastructure for reading and writing, such as post offices, libraries, bookstores, news-stands, cybercafés, etc.). Access, on the other hand, is useful for discussing the social and epistemological conditions necessary for literacy learning and use (see Hernández, this volume). It refers to the social interactions, collaborations, and support for learning discourse practices. In the case of digital technologies, availability refers to installed infrastructure (electricity, devices that work, connectivity, software), and access refers to interaction with more expert others who act as mentors and demonstrate how technologies work, share practitioners' knowledge, and reveal authentic purposes and ways of using the devices.

In a socio-constructivist understanding of learning, the learner is an active participant in the development, appropriation, and structuring of knowledge.

Socio constructivists emphasize the importance of collaboration and interaction with others as an organic part of learning. In everyday contexts, the experts engage the beginners in complete activities rather than fragment them into arbitrary bits based on an abstract notion of what is easy or difficult. Learners must solve complete and very real problems (Lave and Wenger, 1991). In the case of introducing GPS use to the fishers, the more seasoned participants involved the novices in a context where using this particular technology was an important part of their shared activities and its appropriation was a complex process that included understanding its affordances (Sutherland, 2008; see Guerrero and Carvajal, both this volume) and its relationship to other non-digital technologies and fishing practices (Kalman and Liceaga, 2009).

In this chapter, I approach the dissemination of GPS technology and satellite-generated maps in a fishing community as an avenue for understanding how this contributes to fishers' meanings and how their meanings shape their social action. Simply stated, it is a way of understanding why they do the things they do (Geertz, 1983). My conception of the fishers' GPS appropriation privileges the relationship of technology users with others around texts and devices, over the direct relationship between these individuals and the modes of representation they are newly engaging with (see Guerrero, this volume). I am concerned about the social history of how knowledge is constructed, who participates in learning contexts, how opportunities for learning occur, what the broader practices that contextualize learning are, and what learners' purposes for approaching the new are (see Hernández, this volume). Evidently the fishers relate to GPS through reading the screen and registering data, but their knowledge and fluency for understanding, interpreting, and registering data is mediated by researchers and more expert others. Furthermore, it is situated mediation, technology use in specific contexts of use (see Lorenzatti, this volume). Learning to do so is influenced by how the fishers position themselves to the other participants using the technology. This being said, it is important to keep in mind the fishers' use of local knowledge as a resource and, at times, a reference for accepting (or not) new forms of representation, particularly in the case of printed satellite maps of places well known to them.

The data presented here comes from conversations, meeting notes, correspondence between marine scientists and lobster fishers, and my field notes from an exploration of the field as a possible research site. For this reason the commentary regarding the recent fieldwork is tentative in nature while the information derived from the marine resource project is more conclusive.

Lobster Fishing Procedures Before and After GPS Technology

During the 1960s Cuban fishermen showed the fishers of Mexico how to build *casitas cubanas*, which are artificial habitats or refuges that create spaces similar to small caves in the reef (Pérez, 2005). Prior to this technological innovation, the lobster fishers in Mexico dived and hand-picked each lobster. These artificial shelters are approximately 1.50 to 1.20 meters long by 14 centimeters high and currently built out of concrete. The *casitas cubanas* provide a trap for the lobsters, which hide under the cement slabs as if they were natural formations, and the fishers, instead of locating, diving, and picking them up one by one, are able to harvest the lobsters by collecting them from inside the *casitas*. The fishers still have to dive into the water with a *jamo* or net to gather them, but with the artificial habitat the lobsters hide and accumulate beneath the slabs. This technology requires placing the refuges on the sea bottom in rows, similar to the ways farmers grow crops, and returning to them periodically (SCPP Pescadores de Vigia Chico. 2006)

Because of the lobster fishers' use of the artificial habitats, returning to specific points is essential. Once they arrive at the spot indicated on the screen, the fisher and the diver begin to look for the cement refuges under water. Currently, the fishers use a GPS device that has a margin of error of approximately five meters, and they may not stop exactly next to the refuge. Given the dynamics of the waves, once the motor is cut or slowed to an idle, the boat continues to move. Furthermore, artificial shelters that have been in place for a while begin to blend in with the surroundings as they become covered with sand and sea grass. One distinctive characteristic of the *casitas* is their regular rectangular shape set in a background of asymmetrical forms, curves, and textures. Both diver and fisher look over the edge of the boat and on occasion they spot them immediately. When this does not happen, the diver holds on to the edge of the boat with his face in the water and the fisher slowly drags him along until he locates the artificial refuge.

Storms and hurricanes are perhaps the principle cause of refuge displacement, damage, and destruction. After a major storm, fishers have to go to fishing fields and find and retrieve their shelters. Because of their investment, they need to locate and salvage as many as they can, haul them up into the boat, and organize them into rows again. Under these conditions, they have to look for their equipment using traditional means.

Learning to Use GPS

For the duration of their project, the marine scientists travelled Sian Ka'an on a regular basis. Approximately six fishers assisted them with their activities, and they were the first to become familiarized with GPS functioning. As they travelled across the bay, the marine scientists would stand at the back of the boat and give the fisher precise indications where to head the boat, based on their readings of the hand-held GPS device. The fishers had the opportunity to observe the scientists and to ask questions about what the screen displayed and how to set the device. The researchers in turn answered by showing the fishers the screen and interpreting the symbols and maps displayed on it.

The fishers took an immediate interest in the GPS because they saw it as potentially useful for improving their fishing techniques. As an informal situation for learning, these *in situ* demonstrations had an epistemological advantage: the researchers involved the novice fishers in experimental situations, creating on the spot the opportunity to hypothesize about how the GPS works and the immediate conditions to validate or reject their conjectures. When the fishers did not understand what was displayed on the screen or seemed unsure about what to expect, the marine scientists could ask them, "What do you think will happen to this line [on the screen] if we go over there?" They could then redirect the boat with the GPS in hand to see if their expectations were correct or not. This allowed the fishers to experiment and learn by trial and error under the mentorship of the expert GPS users, the marine scientists.

It was not long before several members of the fishing community asked the marine scientists to teach them how to use GPS devices. A two-day workshop was set up and held at the fishing cooperative locale; approximately 40 percent of the town's lobster fishers participated. The course lasted 12 hours; nearly 25 percent of the time was spent in an information session the researchers designed for the fishers. It was very similar to a class at school: the academics projected digital slides and information about the GPS satellite system and introduced specialized vocabulary.

They also projected images of GPS devices and explained how basic GPS tools work. The hand-held GPS device used by the fishers includes buttons marked with abbreviations in English such as **Nav** for navigation; **G** for go to way point; **MOB** for man overboard (used for recording new coordinates).

It also has a small screen where a digitalized schematic map appears as well as a pointer. The GPS operator has the option of seeing the map close up or at a wide angle, a view that shows a larger area but in less detail. The introduction of this technology simultaneously includes language issues, technical procedures, and digital representations of grounds geographically familiar to the fishers.

FIG. 5.1 Fisher holds two different devices.

The rest of the time offered a series of activities using GPS devices to familiarize the fishers with the technology and for them to learn how to record points, set a route, and use different GPS tools while using them to walk around town. The researchers demonstrated how to operate the hand-held device by introducing coordinates of a known site in town. The fishers then had to use the GPS to navigate to that precise location. As the fishers participated in the location activities, the researchers also showed them how to interpret the screen for direction, speed, and proximity to their destination.

Mapping the Bay

In exchange for their help in data gathering on the bay, the researchers introduced another technological innovation into the community by developing maps of the fishing area. The fishers in Punta Allen are organized in a cooperative known as the "*Pescadores de Vigía Chico.*" A particularity of this cooperative is that its members have parceled the bay into *campos* or fields and allotted exploitation rights to its members. In exchange for fishing privileges, the members agree to collectively established rules of self-governance and conservation, such as in what way the fishing area will be divided amongst the members of the cooperative, low quotas, and strict release rules regarding female and young lobsters (Arellano-Guillermo, 2004; SCPP, 2006).

These maps represent 104 of the 120 sea plots and have proved helpful for settling disputes and poaching issues (Sosa-Cordero, Liceaga-Correa, and Seijo, 2008). Previous to the computer-generated maps, the fishers located their fields by using on-shore references, buoys, and underwater landmarks. Fishers are able to make out a variety of shapes in the water, such as turtles, fish, eels, and manta rays, from a distance.

The marine scientists researching the benthic grass project considered the development of these maps an important contribution to the community and referred to them as "*escrituras*," or property titles. They believed the maps had two consequences. First, they gave the fishers precise locations for their plots. In one of our many conversations, one of the scientists noted that, previous to the maps and the GPS technology, the fishers had "their [land] references, their common sense, their sense of direction or whatever." Second, the printed maps made it possible to sell their plots to other members of the cooperative. The marine scientists considered the GPS technology and the maps an improvement over the fishers' existing location practices and local knowledge. To them the maps and coordinates were legitimate knowledge while the everyday practices were not; they seemed to doubt the fishers' abilities to read the light and dark patterns of the sea floor.

Within the community, however, the fishers had a more attenuated response to the new devices and representations. While the GPS devices were used by several fishers, this was not true for all, and some believed that using them was a sign of a lack of more traditional knowledge. A young boy at the local middle school told

FIG. 5.2 Map of 104 individual "campos" or marine plots in Bahía de la Ascensión (SCPP Pescadores de Vigía Chico, 2006).

me that his father did not need a GPS because they "were for fools," for those who did not know how to fish without them. Similarly, only some of the fishers valued their printed maps and used them to keep track of their parcels. Antonio,[5] a well-known local, did not use the maps and in fact did not keep any printouts of them, but knew that the cooperative offices had copies if he required them. Another fisher, however, saw value in them for inheritance purposes; he kept them for his children, so that in case of his demise they would know what he owned. He showed me three maps, of which one had the word *vendido* (sold) on it, and he remarked that he no longer owned that plot but kept the map anyway.

While I was visiting the bay with the marine scientists from Mérida, one of the fishers approached them to request a map of his field. For some reason his areas had been left out of the original mapping (which represents nearly 87 percent of existing fishing fields), and he wanted to have a printed copy of his plot's location, like others in his community. Javier, one of the researchers, sat down with him, showed him the computer-generated image of the bay, and gave him the location of the plot. The marine scientist's explanation was unacceptable to Narjo, the fisher, who argued for a slightly different location for his plot. From here, a tug of war over how to read the map began, the researcher insisting on the precision of the map and the fisher insisting that it was incorrect. He said several times "I don't like that location, I want my plot to be here" and would recall the physical references on the shore that he used to locate his field. He relied on his years of experience fishing on the bay, his knowledge of the landmarks, and how he used them to navigate his boat over the water. The researcher, however, insisted that Narjo's plot had to be in line with certain coordinates and seemed to pay little attention to the fishers' evidence. This was unacceptable to Narjo, who was sure about his field's location. Finally, in exasperation, after repeating several times the landmarks of his plot and naming his neighbors, their plots, and their landmarks, he took a pencil and drew his own representation of the fields. As a way of solving their disagreement, Javier offered to go out with Narjo in a boat and, using a GPS device, set the coordinates and then generate his map. Narjo's confidence in his knowledge of his field's location vis á vis its computer-generated representation is noteworthy; he seemed unwilling to accept another's judgment without going back, looking again, having new first-hand evidence, and participating in the plotting, regardless of the marine scientist's insistence on the precision of his map. It turns out that Narjo was correct to have his doubts: according to the minutes of one of the meetings held by the fishers and the marine scientists, in some cases they detected overlaps in the fishing fields, making it necessary for the coop members to work out their property and fishing rights disputes.

Learning in an Epistemologically Varied Environment

The analysis of how the fishers learned to use GPS devices from the researchers and how satellite-generated maps were appropriated by the community has revealed the co-existence of scientific and everyday knowledge in this fishing community. Through exploring how fishers harvest lobsters and locate their traps, it becomes clear that both types of practices are useful in different situations and one does not completely replace the other. This is also true of interpreting printed maps of the fishing plots. The maps were useful to some as evidence of their fields' locations and their ownership of them and were mobilized in different ways for buying or selling their properties, for inheritance concerns, and for resolving disputes. While the maps and the devices for the most part were valued, acceptance was not homogeneous, and the fishers did not completely relinquish their knowledge of the land or the sea in the face of more standard forms of knowing or representation.

What emerges is an epistemologically varied environment laced with interrelated opportunities for learning activities mediated by others, different tools, and by different interpretative practices and stances. School settings tend to prefer

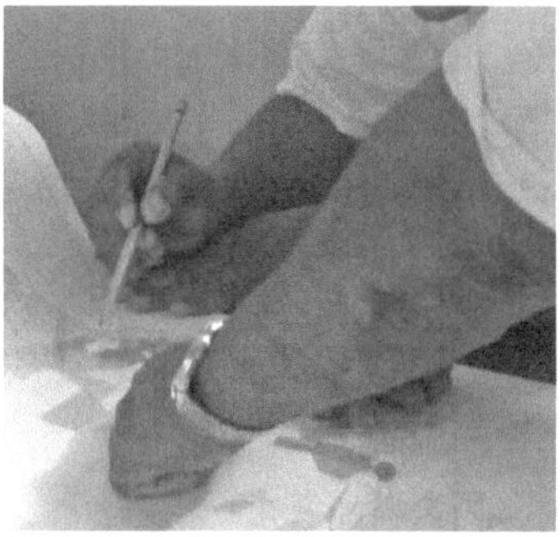

FIG. 5.3

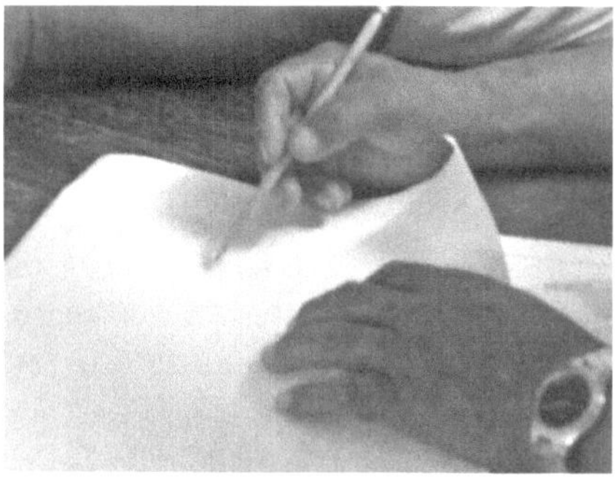

FIG. 5.4

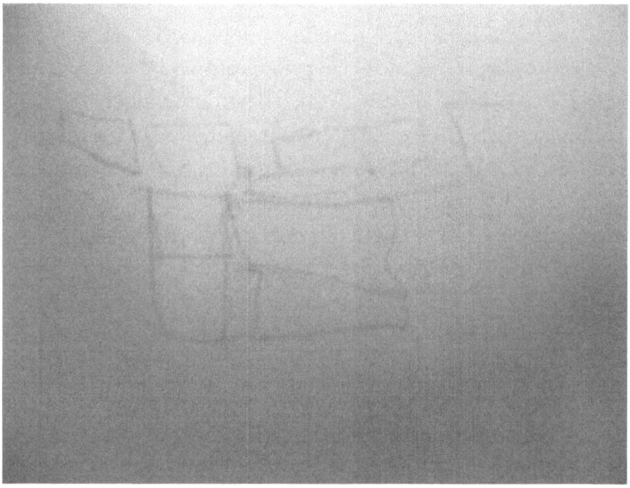

FIG. 5.5

Figs. 5.3, 5.4, and 5.5: Fisher and researcher read digital map of fishing fields together; Fisher draws his own representation.

learning by transmitting information from one participant (teacher) to the others (students) and via pre-defined categories and media, over access to knowledge through direct experience and mediated by social interaction (Packer and Goicoechea, 2000; see Carvajal, this volume). Criteria for selecting and using materials and artifacts during teaching and learning are set by tradition, curriculum, and availability. In contrast, learning between the fishers and researchers—traditional and scientific communities—varies in purpose, format, activity, and outcome.

Through this exploration, mediation and collaboration as key activities for disseminating literacy and technological practices have also become salient. However, while the fishers' access to GPS technology was the product of their interaction with the marine scientists, they were critical in their acceptance of it.

Their opportunities for learning occurred by participating in situated practices, observing how the scientists used the GPS technologies, asking specific questions, and using the devices with the assistance of the researchers. The marine scientists made the technology accessible to the fishers by answering their questions in a variety of ways: they demonstrated the technology use, they explained how to interpret the screens, and they created experimental situations for the learners to learn by a guided process of trial and error. On the boats, these activities were "whole activities" in the sense that demonstrating and experimenting with the devices included reading the screen, using the different functions, and exploring the buttons. The researchers also helped make the devices both accessible and available: they prepared a brief explanation on using GPS, created hands-on situations to rehearse their use, and lent handhelds to the fishers, who could then explore them.

Although in research circles and academic contexts the scientists' technological options and forms of representation are prestigious and dominant, in the fishing community they were questioned, scrutinized, and integrated with other ways of knowing. Evidence such as direct experience, knowledge from years of fishing, and familiarity of local procedures were templates for understanding the technologies brought by scientists and for incorporating them into their repertoires (see Hernández, this volume). These became a new option but did not necessarily replace previous ones. Fishers used their knowledge to decide when and how to use new practices and they determined specific situations in which one was more appropriate, when both were required, or traditional knowledge and know-how was more reliable. They seemed more flexible and willing to consider others' ways of doing things, something the scientists seemed less inclined to do.

From the theoretical perspective presented here, reading, writing, and speaking about literacy—in both print and digital modalities—constitute essential activities of written culture. The presence of the researchers in the fishing community shows the power of well-informed and effective mediators and the potential effects of collaboration for disseminating literacy practices, something I have observed in other contexts as well (Kalman, 2004). The fishers' command of local knowledge for participating in social activities around fishing and field location is an enduring feature of their work. It is enhanced, but not replaced, by using digital GPS devices and digitally generated maps.

Notes

1 Parts of this chapter are based on a paper previously published in *Maritime Studies* (*MAST*): Kalman, J., and Liceaga, A. (2009). "The coexistence of local knowledge and GPS technology: looking for things in the water." In *MAST/Maritime Studies* 8 (2), 9–34. My thanks to the editors at *MAST* for permission to republish parts of that article and to Comunidad y Biodiversidad, A.C. for permission to use map on page 75.

2 On January 20, 1986, Sian Ka'an became part of UNESCO's Man and the Biosphere Program. According to UNESCO, "Biosphere reserves are areas of terrestrial and coastal ecosystems promoting solutions to reconcile the conservation of biodiversity with its

sustainable use. They are internationally recognized, nominated by national governments and remain under sovereign jurisdiction of the states where they are located. Biosphere reserves serve in some ways as 'living laboratories' for testing out and demonstrating integrated management of land, water and biodiversity. Collectively, biosphere reserves form a world network: the World Network of Biosphere Reserves (WNBR). Within this network, exchanges of information, experiences and personnel are facilitated" (UNESCO, 2009).

3 The term benthic refers to "anything associated with or occurring on the bottom of a body of water" —http://oceanservice.noaa.gov/facts/benthic.html.

4 Under the academic leadership of María de los Ángeles Liceaga Correa, from 2000 to 2003 her group collected data for a several environmental variables using remote sensoring technology. See Kalman and Liceaga, 2009.

5 All names are pseudonyms.

References

Arellano, G.A. (2004). Manejo integrado de zona costera y Áreas Naturales Protegidas: La Reserva de la Biósfera Sian Ka'an, Quintana Roo, en:E. Rivera-Arriaga, In E. Rivera Arriaga, G. Villalobos, I. Azuz, F.y Rosado. (eds.), *El Manejo Costero en México*, Campeche: Universidad Autónoma de Campeche, SEMARNAT/Cetys-Universidad de Quintana Roo, 445–54.

Barton, D., and Hamilton, M. (1998). *Local Literacies: Reading and Writting in One Community*. London: Routledge.

Blommaert, J. (2005). *Discourse: A Critical Introduction*. Cambridge: Cambridge University Press.

Collins, J., and Blot, R. (2003). *Literacy and Literacies: Texts, Power and Identity*. Cambridge: Cambridge University Press.

Geertz, C. (1983). *Local Knowledge: Further Essays in Interpretative Anthropology*. New York: Basic Books.

Kalman, J. (2001). Everyday paperwork: literacy practices in the daily life of unschooled and underschooled women in a semi urban community in Mexico City. *Linguistics and Education* 12 (4), 367–91.

Kalman, J. (2004). *Saber lo que es la letra: una experiencia de lecto-escritura con mujeres en Mixquic* [*Discovering Literacy: Access Routes to Written Culture for a Group of Women in Mexico*]. Mexico, D.F.: Secretaría de Educación Pública-UIE-Siglo XXI Editores.

Kalman, J., and Liceaga, A. (2009). The coexistence of local knowledge and GPS technology: looking for things in the water. *MAST/Maritime Studies* 8 (2), 9–34.

Lave, J., and Wenger, E. (1991). *Situated Learning: Legitimate Peripheral Participation*. Cambridge: Cambridge University Press.

Packer, M., and Goicoechea, J. (2000). Sociocultural and constructivist theories of learning: ontology, not just epistemology. *Educational Psychologist* 35 (4), 227–41.

Pahl, K., and Rowsell, J. (2010). *Artifactual Literacies: Every Object Tells a Story*. New York: Teachers College Press.

Pérez, J. (2005). *Parcelan pescadores el mar: cultivan langostas con sombras de concreto* [Fishermen parcel the sea: they cultivate lobsters with shades built out of concrete]. http://www.wwf.org.mx/wwfmex/archivos/am/Access date: January 13, 2009 [Spanish].

SCPP Pescadores de Vigia Chico. 2006. *Sistema de áreas exclusivas para realizar la pesca de langosta* [Exclusive lobster finishing areas system]. Taken from the presentation given at the Pescador a Pescador II Meeting, Laa Paz, Baja California Sur, Mexico,

6–10 de marzo 2006. http://www.pescardorapesacdor.net/pap2/SCPP_VIGIA.pdf. Access date January 24, 2009.

Scribner, S., and Cole, M. (1981). *The Psychology of Literacy*. Cambridge, MA and London: Harvard University Press.

Sosa-Cordero, E., Liceaga-Correa, M. A., and Seijo, S. (2008). The Punta Allen lobster fishery: current status and recent trends. In R. Townsend, R. Shotton, and H. Uchida (eds.), *Case Studies in Fisheries Self-Governance*. FAO Fisheries Technical Paper. Rome: Food and Agriculture Organization of the United Nations, 149–62.

Street, B. (1984). *Literacy in Theory and Practice*. Cambridge: Cambridge University Press.

Street, B. (1995). *Social Literacies: Critical Approaches to Literacy in Development, Ethnography and Education*. London and New York: Longman.

Street, B. (ed.) (1993). *Cross-Cultural Approaches to Literacy*. Cambridge: Cambridge University Press.

Sutherland, R., Robertson, S., and John, P. (2008). *Improving Classroom Learning with ICT*. London: Routledge.

UNESCO (2009). *The World Network of Biospheres*. Available: http://portal.unesco.org/science/en/ev.php Access date: March 8, 2009.

Walker, R., Taylor, J., Waska, H., Ponce-Taylor, D., Vause, B., Comley, J., Visvalingam, S., and Raines, P. (2004). *Sian Ka'an Coral Reef Conservation Project Mexico 2004 – Final Report*. London: Coral Cay Conservation Ltd.

Warschauer, M. Reconceptualizing the Digital Divide. *First Monday* [Online] 7 (7) (1 July 2002), http://firstmonday.org/htbin/cgiwrap/bin/ojs/index.php/fm/article/view/967/888. Accessed January 15, 2008.

6

WHEN ILLITERATE ISN'T ILLITERATE. READING REALITY IN A MULTIMODAL WAY

María del Carmen Lorenzatti

For the last several years the literacy of youth and adults has occupied an important place in discussions held by international organizations and in the agendas of the Latin American region's ministries of education.[1] Underschooled adults and youths are believed to be incapable of having an opinion, understanding complex issues, and participating in social spaces. People tagged and classified as illiterate, uneducated, or poorly educated are generally not recognized as active social subjects despite the fact that they participate in institutions, struggle for their rights, constitute families, or plan and develop projects for the future. For the most part, the ways they appropriate and use written language is unknown to many scholars, educators, or policy makers because the ways they use written language and other semiotic representations is unfamiliar and somewhat unconventional from a literate's point of view, making their meaning-making processes invisible to the everyday eye. In this paper I will argue, contrary to popular belief and official discourse, that people like Marta Graciela, one of the adults I studied, interpret intricate multimodal representations and participate actively and competently in a variety of social contexts. I will present a complex outlook on literacy learning in order to understand how Marta Graciela interprets the meaning of texts, graphic images, and oral language, and how she uses them in daily life. I will also identify the different tools that mediate her acquisition of knowledge.

Various authors have studied underschooled adults and the literacy knowledge they construct and use in their daily life (Kalman 2003, 2004; Niño Murcia, 2004; Zavala, 2002). In my recent work[2] (Lorenzatti, 2009) I have analyzed how adults who never went to school during their childhood or adolescence make sense of everyday situations using different modes of representation present in their daily life (Kress, 2003).

Literacy and Multiple Modalities: The Conceptual Approach to Understanding Knowledge Acquisition Processes in Unschooled Adults

The theoretical perspective of New Literacy Studies (NLS) focuses on literacy as social practice within wider contexts. This approach offers a methodological and ideological position for debating the apparent neutrality of written practices and creates concepts for analyzing the social and cultural diversity in Latin American countries (Kalman and Street, 2009). When the bond between written and oral language is conceived of as more flexible, their relationship is more complex, and thus form and function in real communication overlap in certain situations. The ideological concept of literacy highlights the fact that written language plurality is historically conceived and culturally built and therefore inserted in relationships (Zavala, Niño Murcia, and Ames, 2004). This suggests examining who reads and writes at each event, how reading and writing are accomplished in the different social circles where the subjects move, and how the expectations and consequences related to reading and writing play out.

These meaning-making practices include not only speaking, reading, and writing but also other modes of representation (music, movies, and graphic images), as well as the vital presence of mediating tools, either human or symbolic. Kress's (2003) outline on multimodality offers theoretical possibilities for understanding how adults that did not go to school during childhood and never learned to read written language well manage to build and give meaning to different semiotic representations occurring in their daily lives. The recognition of the existence of numerous cultural resources involved in message construction (oral language, writing, images, gestures, music, and objects as tri-dimensional models) helps us to overcome the dichotomous view of literacy/illiteracy which has long dominated pedagogical discussions and provides new analytical perspectives for studying the meaning-making processes that emerge in situations of complex graphic compositions (see Carvajal, this volume). Kress posits that the written mode will be increasingly displaced by image in many areas of public communication, even though written language continues to be preferred by the cultural and political "elites."

Accepting that writing is but one of the modes of representation present in texts (Kress and Bezemer, 2009) could potentially influence (and renew) the design of educational opportunities for unschooled and underschooled adults. Kress points out that writing and image are based on different logics of representation and clearly have different affordances. This term refers to the potential of the different modal components that make appropriations and different significations possible.

Written language is generally organized in a linear way and read in one direction, from left to right in Romanic script; for example, the setting of graphemes and spaces form words. The reader must fill the words with meaning, constructing from context, and polysemic options and cultural contents. On the other hand, the organization of image is controlled by a spatial logic, and by simultaneous representation of its visual elements arranged in a spatial disposition. Here the observer's role is to give order to the simultaneous elements.

At present there is a strong multimodal component in different environments and it is important to examine the affordances of these different modes of representation. This approach allows the analyst to reconstruct the ways adults like Marta Graciela interact with socially constructed signs without centering on writing but by adapting it to other modes (Kress, 2003).

Images, Flavors, and Fragrances. Marta Graciela: A Case Study

It is important to understand Marta Graciela's history in order to appreciate some of her living conditions and their significance for the argument concerning literacy and multimodality that I will pursue in this chapter. She was born in a rural area of Bolivia where only boys attended school—girls were not allowed to—and as a result only men learned to read and write in formal settings. Since she did not go to school, Marta felt obliged to learn how to use other forms of representation, such as oral language, images, and gestures according to the communicative practices that arose from the diverse contexts and activities that made up her life. In order to investigate these activities and do justice to Marta's communicative skills, I accompanied Marta Graciela on several occasions to different sites in Córdoba (Argentina) to observe how she dealt with the demands for reading and writing she encountered during her day. I also interviewed her about specific situations. When possible I audio recorded and took photographs as well as extensive notes. I draw upon this material below in my account of Marta's engagement with literacy and other modes.

Although she can sound out some words, she depends heavily on other elements carrying meaning in order to understand the texts she encounters during her day. Images and oral language were the modes most used by Marta Graciela in order to become part of and participate in different social circles. For analytical purposes, the different ways she makes meaning will be analyzed separately, although in her daily activities they are simultaneously present in most cases.

Marta Graciela recognizes letters, numbers, and written words in different media. At night school she works on traditional school exercises like copying words and making sentences, but she struggles to sound out words, find the right letters, and make sense of what she is asked to do. She often waves her hand in the air and calls out "Teacher, Teacher!", trying to get the instructor's attention and help. In this chapter I will portray a different Marta Graciela and illustrate how in her daily life she makes use of other modes and cues, such as graphic image, flavors, and fragrances. In the interest of showing her to be a competent meaning maker, I show how Marta Graciela is a sophisticated reader of multimodal texts even though she flounders when asked to pronounce words or read a sentence out loud in class. There is also evidence to suspect that in other contexts Marta Graciela grasps the meaning of writing, although she is reluctant to admit this openly.

Image is one of the modes present in Marta Graciela's social spaces that make up her daily contexts. Photographs and drawings are found on religious objects, on posters in shop windows, on domestic tools, in different institutions on the street, and so on. What follows is a description of some of the situations Marta Graciela experienced daily where graphic images contributed directly to her understanding of meanings. I will also examine how Marta Graciela uses images in religious worship, her household, and public institutions. In the context of her religious practice, she interacts with different texts containing images. She remembers that when she was an assistant teacher of a group of students she used religious magazines and a Bible for children, which had many colored and black and white images.

In the illustrated version of the Bible, each verse is accompanied by several drawings allusive to its content to help people understand the pastors' sermons and facilitate the memorization of the verses. At an evangelic group meeting I attended, Marta Graciela asked for the reading of a verse she had identified from an image and associated with a situation she was living at the time.

There are also full-page illustrations that emphasize the main actions of each prayer. Marta Graciela uses these images as a reference for her interpretation of the text; for example, she refers to an illustration to analyze the presence of divine intervention in human life and relates it to a mother's care. Looking at the picture, she muses: "This is a bird feeding its little birds. This is how God looks after us, like this bird takes care of its young. I mean, kids are baby birds and mothers look after them" (Registered 4/10/2007).

The presence of images is not only important in the religious domain but also in the domain of work. Marta Graciela sells cosmetics from a catalogue and the drawings and photographs used to illustrate it help her sell the products. She goes through the neighborhood, door by door, offering products to people. She also does this with the people that attend adult school with her. To sell her products, first of all she shows potential customers the catalogue. She asks them to look at the products available, especially those which are on sale at a discounted price. Then, when her customers choose a product, she asks them to write their names on the page, on top of the image they want to buy. The writing is done by the buyer and serves as a reminder to Marta Graciela, but she also memorizes the chosen product. This same procedure is repeated when she places her order. She asks a friend of hers to fill out the order form using the information from the catalogue. Finally, when the products arrive, she organizes them according to her clients' previous requests using their names handwritten in the catalogue to guide how she packages them.

All these actions involve a graphic image and the use of her memory. Marta Graciela identifies the product by its photograph and then remembers the corresponding product for each client. During my interviews with her she did not clearly say if she could read the client's name. When I insisted, and asked her this question again, she simply replied: "I remember it." She completed the final job leading up to the distribution of the products to her customers at home. There she would place the products on the table, sort them into plastic bags, and give them to each client, depending once again on her memory.

At home, Marta Graciela also distinguishes written objects using graphic images. Due to her poor health, she takes several medications, and controls closely what she eats. How she identifies the pills she has been prescribed gives a glimpse at how she has appropriated literacy and multimodal practices. For example, she must differentiate the pills she takes for her cough from those she takes for gastritis. My questions about how she does this led to a description of each medicine and clearly showed how Marta Graciela closely paid attention to the different characteristics of her medications: she described them in detail regarding their shape, color, and size.

Marta Graciela also drinks several different kinds of medicinal teas and she uses color to recognize and distinguish the packages. For example, a yellow box means black tea; green and white indicates a digestive tea, which she also identifies with the image of the plant's leaf that is shown on the outside of one of the boxes. Although she considers the graphic image and the color the most important characteristics, she finally adds the sense of smell and the identification of aromas as a new indicator as well. She associates the plant drawn on the box with a real leaf and, as a result, she succeeds in giving a particular meaning to the box and recognizes the type of tea. Another situation in which she uses images is when she is in the supermarket. Here she interprets different characteristics of the merchandise to distinguish different products from each other. During a visit to one such store, I saw how she used the designs of commercial wrappings, colors, and smells. At one moment, she chose two bars of soap, and, holding one in each hand, she asked me if I could tell the difference between them, placing me in the position of the person with "reading" difficulties that literacy programs often assume for their learners (Rogers, 2005). It was not easy, since they were quite similar. As she sensed my hesitation, she started giving me some hints to help me identify them. "This has one color [lifting one bar of soap], and this one has another [as she lifted the other one]. Yes, yes, here … if you take a look [she smells the soap]. This one must be herbal or something like this. What does it mean? You should have a look at the colors. They mean something. That's the way life is" (Reg. September 12, 2007).[3]

Because the wrappers and their designs were quite similar, Marta Graciela chose to give priority to the color of each one. To this she added another characteristic of the product, its smell, and this helped complete the meaning. In fact, the soaps did have different fragrances. Marta Graciela made explicit the relationship between these modes of interpretation and knowledge. "Let's see, what about this one, what does it smell like? Let's see … this does not smell very much, let's see this one, it's more perfumed. Sure, this is the way to learn!" (Registered 09/12/2007)[4].

These examples illustrate that reading is a lot more than deciphering a chain of letters. In this particular case, where consumers must differentiate products before purchasing them, Marta Graciela interpreted several aspects of the goods in order to make a decision. It illustrates how the product's characteristics—colors, flavors, and fragrances—allow her to pay close attention to the objects she picks up and compare them. This kind of multimodal interpretation is used by reading customers—it is not only a strategy used by non-reading buyers. Readers also factor in and use different representative modes to give significance to their actions: for example, when purchasing a product, even if it has not expired, a customer can determine that it is rancid by the way it looks and smells and prefers not to purchase it. These different ways of reading reality integrate text, image, smell, color to show how complex messages are in a multimodal sense. Letters are not the only carriers of meaning when dealing with physical objects—color, design, smell, size, and shape work together to create meaning and it is the meaning-maker's job to interpret them.

The example of the hand soaps is a complement to another situation in the supermarket. Marta resorted to multimodal representation to recognize a type of laundry detergent when a woman shopper in the store was looking for a low-foaming soap for her automatic washing machine. The shopper first asked me where the low-suds detergent was and I immediately tried to read the labels on different packages to locate it. This was somewhat time consuming, so Marta Graciela told me, "Look over there," pointing to the bubbles logotype that identified the low-foaming product. In this case the image of the bubbles was enough to choose the product. The existence of central and lateral elements signals readers to shift their sight from the center to the sides of the packages. This example illustrates how the image itself and its elements are full of meaning and that it is not necessary to decode every written word to interpret the commercial packaging (Kress, 2003).

Another telling example is how Marta Graciela situates herself in the city. The urban landscape makes it possible to travel through the streets without having to decode the signs. During my time with Marta Graciela, I could see how she used the shapes and colors of signs, landmarks, and buildings to orient herself and at a downtown corner she told me she recognized a certain crossing from the big red and white sign of a well-known transnational company on a specific corner. She moved around the city center by identifying signs and buildings at each corner instead of reading specific street names.

In a previous paper (Lorenzatti, 2007) I described how a group of youths who attended night school identified commercial firms from their signs. While going through the neighborhood, these youths could express their knowledge about the consequences of written culture. They saw that some shops did not post any written commercial identification, which they interpreted to mean that the owners might be evading taxes. In this example, these youngsters are aware of their social world, the obligation to pay taxes to the State, and its close relationship with displaying written identifiers (or not). Apart from being guided by the physical landscape of the city, Marta Graciela is also able to read some signs. Like in any other reading event, this identification is associated with the context where the reading was taking place. For instance, I took her to visit a lawyer to help her with a legal procedure needed to get the deed to her house. When we were looking for a place to park the car, she saw a parking lot halfway down the block. I asked her how she had recognized it and finally she said she could read a little. From the different situations I observed her in, I concluded that she could read some texts, and also used writing to make interpretations, although it was never clear to me how she used them (Nabi, Rogers, and Street, 2009, on "hidden literacies").

Marta Graciela uses images to help her with her daily chores in a number of ways: since she does her shopping in the center of the city, she uses public transportation to get to shops and stores; she is able to identify goods in supermarkets and neighborhood shops using images; and she also uses a catalogue to sell cosmetics. In church Marta Graciela works as a teacher's aide at Sunday school and as part of her duties she uses pictures to teach children how to interpret Bible stories. The examples examined here show Marta Graciela to be an active, engaged interpreter who makes meaning from a variety of representations. Her signifying work is a far cry from the usual passive social portrayal of illiterate adults as dependent and limited.

Oral Language and Mediation

Much of the spoken interaction that occurs in the evangelic church she attends in her neighborhood is based on the written word and this context provides a powerful illustration of how talk plays a fundamental role in the interpretative practices of written texts. At church meetings, the leader's oral word relies on counseling and interpretation of the written word. The spoken word is used to mediate God's speech, as represented in writing in the Bible. The only person in the meeting who has the sacred text in front of her is a female leader. She is the one who reads it out loud and interprets what it says; as she reads the verses she also explains the general meaning of the text as she understands it. Her speech fosters the participation of others, who give significance to the text based on their life experiences. The congregation exchanges dialogues, shares points of view on any given problem, and offers advice and counsel to each other on different aspects of their personal lives.

They believe that all actions will always be guided by the light of the divine word. Marta Graciela says: "what is important is to read the word which gives food to our souls. Greatness is to know God's word, this word makes you free. Let us say the word of God will heal you, here is health, and here is the word." The vision of the word is regarded as a sign; a way of influencing others and oneself. Here the word, written and oral, is viewed as a sign from God and has influence over people's actions. Zavala (2002, 176) points out that "the word can leave the page so as to irradiate different forms of power." From what I witnessed at the church meetings in this barrio in Córdoba and Marta Graciela's expressions during the interviews, for her, the Bible is a written document that reveals the "truth," guiding the faithful in their daily actions. In this particular religious context, it is the spoken word that sustains social practices in the written culture.

Marta Graciela's approach to spoken language in the other institutions we visited differed from how she understood it at church. In other social spaces, particularly in large institutional settings such as public agencies and banks, she interacted with employees as a first-hand strategy for navigating the bureaucracy and institutional procedures. In these institutions oral language many times became a bridge towards achieving particular goals, such as collecting public support funds, paying a bill, having a teller read to her from the computer screen. For example, when I accompanied her to the electric company, she asked numerous questions of the employees until she reached the office where she had to file a complaint about being overcharged. On entering the building she greeted the guard and informed him of the reason for her visit; she asked him then where she should go. He escorted her to the billing office. On the way there she again asked a cashier about which way to go, albeit in a different way. This time she tried to explain her situation so she would be sent to the precise office and the person in charge. Furthermore, the cashier did not give the information verbally, she answered with gestures, which were understood by Marta Graciela. However, she determined that this information was not precise enough and asked yet another person. In this case it was another security guard to whom she directly made the question: "Where is 'billing'?"

The sequence of questions and answers she pursued and the dialogues with different people once inside the institution gave her the necessary information to get to the right office. In this particular example she uses the interrogative form as a way of finding her way physically through the institution and it also works as a mediator of the written signs posted along the way to indicate the locations of different offices.

Gestures, looks, and common shared meanings in different social contexts form part of the conversational knowledge all speakers use, regardless of their fluency as readers or writers (Gumperz, 1984). Based on the evidence presented here, it is possible to argue that Marta Graciela is not handicapped, nor dependent, nor incompetent: she navigates through the institutional landscape in ways similar to how others do it. The difference is that throughout her life she learned to conjugate diverse sources of information, and to ask for assistance when she needed it, not to depend exclusively on the written word.

Asking questions to get directions is not the only way Marta Graciela uses interrogative forms. On other occasions she asks questions, but in this case it is to confirm what she already knows. I observed this when she collected her state aid pay. Instead of standing in line, she went right up to the bank teller and asked a series of questions in order to get more information, because she does not trust the department of labor data (the department of labor authorizes her stipend). Marta Graciela moved towards the cashier to confirm the right place to collect her money: "Are you paying for the plan P?" The cashier did not give her a direct response, he only answered: "You have to line up."[5] Then she asked a general question about her situation. Since she got an unexpected answer, she changed the focus of her question by adding: "I do not have to wait" (September 20, 2007), indicating her specific purpose of getting near the counter. Marta Graciela seeks confirmation of her ideas so as not to spend more time at this place than necessary, particularly if she is not going to be able to pick up her stipend there. Since she often invests a lot of time in doing this type of procedure, she dislikes queuing and not getting positive results, and has developed this strategy to shorten her waits.

Marta Graciela also asks people to help her with her own writing needs. She spent a lot of time doing paperwork and tried to keep a record of the offices she visited. She explained, for example, how she located office buildings she visited for the first time and how she solicited help from strangers, asking them to write down the exact address for her. On one occasion, she had to go to an agency located downtown and a neighbor had given her directions on how to get there, but she did not have the street number. She took the bus to the correct street and got off at the stop nearest the location of the building. The following excerpt from her narrative explains the procedure she followed for keeping a written record of the address:

> What do I have to do? I clearly point to the right place now, they write it down for me … I pay attention to the surrounding setting … to the way I should go … I get off the bus on the corner for example and I have to move back … yes. **Yes, that is the way I have managed for my whole life.**[6] (Registered September 20, 2007)

Once in the vicinity of the building, Marta Graciela gets off the bus and walks back towards the approximate location of the office she wants to visit and looks around to get her bearings. Once she locates the building, she asks a person to write down its exact address on a piece of paper she carries for this purpose, or on the side of a printed document. She points to the building and the passerby writes it down so that she can keep a record of her visit and have the address in case she has to return. She also makes a mental note of surrounding landmarks such as billboards or store fronts.

Marta Graciela is conscious of her relation with mediators and the role they often play in her encounters with the bureaucracy. She states, "There is always someone who knows," signaling her ability to find out what she needs to know by mobilizing the knowledge of others. Relating to others socially for Marta

Graciela involves questioning and re-questioning, a strategy she refers to as "making life." She uses others' presence and voices to confirm what she already knows. As suggested by several authors, Marta Graciela uses oral and written language in deliberate ways to establish and maintain social relations and to participate in social and culturally valued activities (Dyson, 1997; Heath, 1983; Kalman, 2004).

Mediated Multimodality

Thus far I have argued that Marta Graciela interprets complex representations and participates actively in a variety of social contexts that require interaction with others around texts. I have illustrated through a series of examples how she used visual image and oral language to acquire knowledge, and interpret written objects in different social spaces. However, as illustrated by the example of keeping her own written records, these representative modes do not operate independently; they operate in synchronization with the participation of other people who become brokers or intermediaries for Marta Graciela.

Relevant research has concluded that the appropriation of knowledge takes place during collective activities with others—in the case of literacy, this alludes particularly to other readers and writers (Kalman, 2009; Lave and Wenger, 1991; Rogoff, 1999). Multimodality per se does not necessarily generate appropriation, but it does provide coinciding cultural resources in the construction of meaning. Marta Graciela simultaneously interacted with different modes of representation and mediators as a way of constructing needed situated meanings.

One of the places where intermediaries were abundant was at her church. Here, different people act as mediators of the divine word. They interacted with Marta Graciela in different ways: the pastor, the church group leader, and some neighbors who are part of the congregation and also Marta Graciela's friends all predict and share the word of the Bible with a group of persons. For example, the pastor—a woman long active in the church—gives sermons and teaches at the Biblical Institute. Marta Graciela attends both places. The pastor explains the "holy word" and promotes the participation of the faithful by creating a closer and more direct approach to the verses and by inviting the women attending the meeting to express their thoughts and share their personal problems. In some situations Marta Graciela resorts to others' assistance; she may need a person to confirm what the pastor has said in order to fully understand and remember the verses. Sometimes Marta Graciela meets with some of her neighbors at her home to read the Bible, and in this context she becomes the preacher.

Intermediates also participate in the work environment; they read commercial catalogues to Marta Graciela to indicate who is buying the wares, or the price of the products. She accepts this relation with "the others" who she believes know more than she does; that is why she demands their guidance. In her close circle of neighbors, friends, and family there is always a person who can help guide her.

This is easily seen when she says: "There is always someone at home to whom I can call for help, sometimes the clients themselves."

In public institutions she not only interrogates the employees about what she does not know but also decides who she needs to interrogate. The questions are directed to those she thinks will have the answers she needs and will be able to help her, fundamentally because of their position in the institution, such as a security guard or the bank teller. It appears that Marta Graciela's choices regarding whom to talk to or who might mediate written text for her are not random choices. She selects them according to her knowledge about institutions and employees who will have access to the information she needs.

This evidence shows how Marta Graciela actively develops "mediated multimodal strategies"; she incorporates knowledge about social contexts, issues, and situations, and uses literacy by integrating multiple modes of representation as well as mobilizing the participation of other people. When recognizing her own strategies Marta Graciela considers herself to be an intelligent person because she is able to successfully invite other people to read to her and in this way she is able to satisfy the social demand for dealing with written texts. Interacting with others to convey meanings has proved to be a common daily activity (Heath, 1983).

Kalman (2003) reviewed research from several sources on mediated literacy practices from several countries. One of these papers (Scheffelin and Cochran-Smith, 1984) describes the role immigrants' children play as intermediaries between the adult family members and schools, international organizations, and companies. Another study shows how working-class adults ask for help from neighbors and more literate friends to meet their reading and writing needs (Fingeret, 1983). Asking for help did not make these adults feel dependent on others or ashamed. According to the evidence presented in this chapter, Marta Graciela did not feel inferior to others when she required help to guide her actions either.

Final Thoughts: The Social Distribution of Literacy and Mediated Survival

Several years ago Scribner and Cole (1981) defined literacy as a social practice and they considered it as a sequence of activities that require the use of technology (pencil or computer), abilities (specific knowledge—"know how"), and social knowledge. Street (1995) broadened this definition to include what people think or believe about reading and writing, not only their observable actions.

In Marta Graciela's case, literacy can be understood in relation to her use of social knowledge in order to meet demands for reading and writing rather than her personal control or mastery of written language. It is also important to point out that she selects and uses technology according to a given situation. For example, she uses a pen to sign papers at the bank; she explores product packages and recognizes product designs, and uses images to sell cosmetics; she points to the computer screen in the bank. She also makes deliberate choices about how to approach a given social demand for reading and writing while simultaneously shaping her use of oral language. For example, in institutions she uses questioning as a way of navigating her way through the building to find a specific office, and in church the word takes on the value of sole and unique truth. She also understands certain types of printed material. She uses many devices and texts in her interactions with others and picks up on their forms of expression.

This suggests that Marta Graciela's unfamiliarity with many aspects of written language is not attributable to a lack of individual capacity or interest in learning to read and write, but to the inequalities inherent in the social distribution of literacy. The evidence shown throughout this chapter provides an empirical basis to discuss how multimodality plays a role in individuals' lives. The convergence of the multiple modes of representation (oral language, writing, graphic images), the presence of mediating tools, and people who act as mediators are at the center of how Marta Graciela makes meaning in the social and symbolic world. Including the interpretation of multimodal resources adds a new dimension to the definition of literacy. Being a reader or writer implies using current technology and deciding on what aspects of reading and writing one wishes to accomplish (Meek, 2004). It also implies that the presence of mediators is part of how social literacy practices work and how interpreting and producing a written text is related to the language practices that appear in the context of use (Kalman, 2003). Finally, this study displays the different ways people with reading and writing experiences similar to Marta Graciela's make use of varied modes and strategies to understand texts that appear in daily life.

Notes

1 For example, The Ibero-American Programs of Literacy and Basic Education of young and adult individuals approved at the XVI Ibero-American Ministers of Education Conference (Uruguay, 2006). The resulting document shows progress in some of the concepts related to most national literacy programs. Its authors also consider that these problematic issues should be transformed in a state policy and included in a wider process of basic education. Short-term campaign models associated with political opportunities should be avoided. It is also necessary to connect literacy with other social programs, like income creation, health, and basic sanitation, among others.

2 The research work was developed between 2005 and 2009 from a socio-anthropological perspective. Its central purpose was to study the literacy practices of non-schooled adults and their social environments: the church, public institutions (bank entities, Labour Ministry, electric lighting office services, school for young and adult individuals), their own places, a legal office, a tea house, and the urban sector. Semiotic and sociological approaches contributed to understanding some of the ways they construct knowledge and appropriate communicative practices as well as their influence in their creation of a personal project for the future. Attention was paid to understanding the role that school plays in these adults' lives.

3 *"Este tiene un color [levantando el producto con una mano], éste tiene otro color [haciendo lo mismo con la otra mano]. Si, si Ud. se fija ... mirá [lo huele]. Éste debe ser de algo de hierbas o algo así y ¿qué significa? ... Ud. se tiene que fijar por los colores, significa una cosa. Así es la vida."*

4 *"A ver y éste ¿qué olor tiene? a ver ... no tiene gusto a tanto ... a ver ... es más perfumado. Claro ¡¡así se aprende!!"*

5 *"Están pagando el plan P?", el cajero no le contestó de manera directa: "Tiene que hacer la cola."*

6 ¿Qué tengo que hacer? Señalo bien un lugar, me anotan ... esa vez lo señalo bien, miro que hay al frente ... cómo tengo que hacer entonces ya me fijo ... me bajo por ejemplo en la esquina y me tengo que volver ... si. *Si, así, así me he manejado toda la vida*" (Registered, September 9, 2007).

References

Dyson, A. (1997). *Writing Superheroes: Contemporary Childhood, Popular Culture, and Classroom Literacy*. New York: Teachers College Press.

Fingeret, H. (1983). *Adult Literacy Education: Current and Future Directions. An Update.* Columbus, OH: ERIC Clearinghouse.

Gumperz, J. (1984). Introduction: Language and Communication of Social Identity. In J. Gumperz, *Language and Social Identity*. Cambridge: Cambridge University Press, 1–21.

Heath, S. (1983). What no bedtime story means: narrative skills at home and school. *Language in Society* 11, 49–72.

Kalman, J. (2003). *Escribir en la plaza* [*Writing on the Plaza*]. México, D.F.: Fondo de Cultura Económica.

Kalman, J. (2004). *Saber lo que es la letra: una experiencia de lecto-escritura con mujeres de Mixquic* [*Discovering Literacy: Access Routes to Written Culture for a Group of Women in Mexico*]. México, D.F.: Secretaría de Educación Pública-UIE-Siglo XXI Editores.

Kalman, J. (2009). Literacy partnerships: access to reading and writing through mediation. In K. Basu, B. Maddox, and A. Robinson-Pant, *Interdisciplinary Approaches to Literacy and Development*. London: Routledge, 165–78.

Kalman, J., and Street, B. (eds.) (2009). *Lectura, Escritura y Matemáticas Como Prácticas Sociales: Diálogos con América Latina* [*Reading, Writing and Mathematics as Social Practices: Dialogues With Latin America*]. Mexico City: Siglo XXI Editores and CREFAL.

Kress, G. (2003). *Literacy in the New Media Age*. London: Routledge.

Kress, G. and Bezemer, J. (2009). Escribir en un mundo de representación multimodal [To write in a world of multimodal representation]. In J. Kalman and B. Street (eds.), *Lectura, escritura y matemáticas como prácticas sociales: Diálogos con América Latina.* [*Reading, Writing and Mathematics as Social Practices: Dialogues with Latin America*]. Mexico City: Siglo XXI Editores and CREFAL, 64–83.

Lave, J., and Wenger, E. (1991). *Situated Learning: Legitimate Peripheral Participation*. Cambridge: Cambridge University Press.

Lorenzatti, M. del C. (2007). *Saberes y conocimientos acerca de la cultura escrita: un trabajo con maestros de jóvenes y adultos* [*Know How and Knowledge about Written Culture: A Project with Teachers of Young People and Adults*]. Córdoba, Argentina: Imprenta de la Facultad de Filosofía y Humanidades, Universidad Nacional de Córdoba Argentina.

Lorenzatti, M. del C. (2009). *Conocimientos, prácticas sociales y usos escolares de la cultura escrita de adultos de baja escolaridad [Knowledge, social practices and school uses of adults' written culture/literacy with a low level of education]. Doctoral Dissertation*, Universidad Nacional de Córdoba, Argentina.

Meek, M. (2004). *En torno a la cultura escrita* [*On Being Literate*], Rafael Segovia Albán, trans. Mexico City: Fondo de Cultura Económica (original work published 1991).

Nabi, R., Rogers, A., and Street, B. (2009). *Hidden Literacies: Ethnographic Studies of Literacy and Numeracy Practices in Pakistan*. Bury St. Edmunds: Uppingham Press.

Niño Murcia, M. (2004). "Papelito manda": la literacidad en una comunidad campesina de Huarochirí ["Paper talks": literacy in a rural community of Huarochirí]. In V. Zavala, M. Niño Murcia, and P. Ames, (eds.), *Escritura y Sociedad: nuevas perspectivas teóricas y etnográficas* [*Writing and Society: New Theoretical and Ethnographic Perspectives*]. Lima, Red para el Desarrollo de las Ciencias Sociales en el Perú [Network for the Development of Social Sciences in Peru], 347–65

Rogers, Alan. (2005). *Non-Formal Education: Flexible Schooling or Participatory Education?* New York: Springer Science+Business Media.

Rogoff, B. (1999). *Aprendices del Pensamiento: el desarrollo cognitivo en el contexto social* [*Apprenticeship in Thinking: Cognitive Development in Social Context*], Pilar Lacasa, trans. Buenos Aires, Barcelona, España: Editorial Paidós.

Schieffelin, B., and Cochran-Smith, M. (1984). Learning to read culturally: literacy before schooling. In H. Goelman, A. A. Oberg, and Frank Smith (eds.), *Awakening to Literacy*. Portsmouth, NH: Heinemann, 3–23.

Scribner, S., and Cole, M. (1981). *The Psychology of Literacy*. Cambridge, MA: Harvard University Press.

Street, B. (1995). *Social Literacies: Critical Approaches to Literacy in Development, Ethnography and Education*, Vol. 7. London and New York: Longman.

Zavala, V. (2002). *(Des) encuentros con la escritura: escuela y comunidad en los Andes peruanos* [*Mishaps with Writing: School and Community in the Peruvian Andes*]. Lima, Perú: Red para el Desarrollo de las Ciencias Sociales en el Perú.

Zavala, V., Niño Murcia, M., and Ames, P. (eds.) (2004). *Escritura y Sociedad: nuevas perspectivas teóricas y etnográficas* [*Writing and Society: New Theoretical and Ethnographic Perspectives*]. Lima: Red para el Desarrollo de las Ciencias Sociales en el Perú [Network for the Development of Social Sciences in Peru].

7

INDEXICAL SIGNS WITHIN LOCAL AND GLOBAL CONTEXTS: CASE STUDIES OF CHANGES IN LITERACY PRACTICES ACROSS GENERATIONS OF WORKING CLASS FAMILIES IN BRAZIL

Maria Lucía Castanheira

This text presents data from ethnographic research on writing involving working-class families that live in a neighborhood called Trombetas[1] in the city of Ibirité, which is part of the commuter belt of Belo Horizonte, Brazil. This study contrasts intergenerational data from two case studies of writing in these families,[2] calling upon the concept of indexical signs as a way of analyzing such intergenerational changes in both local and global terms. In the first case study, developed in 1988 and 1989 (Castanheira, 1991), I examined the ways in which children from these families had access to writing *in* and *out of* school. The second case study was initiated in March 2009 after identifying the current addresses of families that participated in the previous research: it aims to examine the consequences of the social changes that took place in the last decades in shaping the ways in which participants use writing *in* and *out of* school and the relationships between these uses. In the last two years, I have had the opportunity to interview members from two families that participated in the previous case study.[3] The current project makes it possible to compare the living conditions across three generations of these families: children interviewed before are now adults, have children of their own, and, in some cases, live with or close to their parents. Contrasting data from these two studies, I aim to identify changes that have taken place over the years, and to examine the implications of these changes for the meanings of literacy and the changing relations of local and global for such people.

An Ethnographic Perspective as a Logic of Inquiry for Investigating Literacy Practices

In both case studies, I adopt a similar research approach: a social perspective on literacy (Barton, Hamilton, and Ivanic, 2000; Castanheira, Crawford, Dixon, and

Green, 2001; Street, 1984, 2005), and an ethnographic perspective (Green and Bloome, 1998; Heath and Street, 2008) to engage in participant observation and data analysis. As argued by Green, Dixon, and Zaharlick (2001), when exploring an ethnographic approach, the researcher engages in the development of *a logic of inquiry* in the process of collecting, analyzing, and representing data. In this perspective, the construction of *a logic of inquiry* is intrinsically tied to the responsive nature of ethnography and, as Bloome (2005) suggests, such "inquiry on language and literacy is defined in the doing and cannot be pre-determined" (276).

In the current study, the construction of its logic of inquiry was triggered by the fact that I was returning to Trombetas after 20 years and by the very nature of ethnographic work, I was engaged with a constant comparative (Heath and Street, 2008) of the present with the past, of one social group or situation with another, and the examination of local and global relationships (Baynham and Prinsloo, 2008; Brandt and Clinton, 2002; Reder and Davila, 2005; Street, 2005). The consideration of these aspects led to the proposition of the following research questions: what is happening to writing (and reading) in this neighborhood? What *changes* can be identified in the way research participants engage in literacy events or literacy practices (Street, 1984) taking place in Trombetas? How does *what is going on* in Trombetas relate to *what is going on* elsewhere? How did what is going on now come into being? What are the implications of such a case study for our understanding more generally of literacy and change?

Identifying and Interpreting Indexical Signs of Changing

In the process of addressing these questions, I became aware of how my inferences about *what is going on in Trombetas* were grounded in various sources of *indexical signs* of change (Wortham, 2001). It is argued by Gumperz (1982) that relevant aspects of a context are used by participants to make inferences about *what is going on* in a particular context and to guide their participation within a group. According to Silverstein (2006), indexical signs can be seen as cues explored by the participants to make inferences and to ground their understandings of the context in which they participate. Indexicality, then, is a phenomenon far broader than language, which, independently of interpretation, points to something—such as smoke (an index of fire) or a pointing finger—that works indexically for interpretation. Social indexicality in the human realm has been regarded as including any sign (clothing, speech variety, table manners) that points to, and helps create, social identity. In the present case, as we will see, the presence of churches and of written signs can be taken to index aspects of social change. The concept applies not only to the participants in this situation but also to the researcher who engages in the same process of interpreting cues or indexical signs to construct his or her understanding of *what is going on* in a particular research setting.

In the case of this study, it was through contrasting different sources of information that I could identify indexical signs of change: I contrasted Trombetas' current landscape with my memories of the past; pictures and field notes and interviews from both case studies. As an outsider (re)entering the research field, I based my initial inferences about changes from rural to urban economy, for example, in the observation of an expressive presence of written signs in Trombetas' streets nowadays against my memories that I rarely saw written signs on Trombetas' streets many years ago. Also, the massive presence of neo-Pentecostal churches in the area in 2009, sometimes three or four of them in two blocks, in contrast to the absence of any kind of church twenty years ago, were taken as indexing changes with respect to the role of religious practices in this neighborhood. Initially, this exploration of the constant comparative ethnographic principle (Heath and Street, 2008) in identifying and interpreting indexical signs of changes in Trombetas led to the identification of changes in two domains: economic and religious.

The contrast between research materials from both case studies that consist of notes, interview transcripts, and pictures produced in 1988/1989 and 2009/2010 became a source for confirming or questioning my initial inferences. This kind of contrast has been guided by the following question: what changes can be identified by contrasting pictures, field notes, and interview transcripts from 1988/1989 and those produced in 2009/2010?

During interviews, changes in economic and religious literacy domains previously identified were talked about by interviewees, giving access to how they related to those changes (e.g., how they would talk about the presence of churches, what they would do with religious texts). Thus, interviews made it possible to gain different access to what people do with texts, how they position themselves in relation to texts, and what texts do to them (Barton, Hamilton, and Ivanic, 2000) in the economic and religious domains.

During interviews, it was also possible to learn about aspects of interviewees' lives related to home/school literacy, such as the way people reacted to new educational policies, free distribution of textbooks for students, and to the conceptual and material characteristics of these literacy artifacts (Pahl and Rowsell, 2010). So, within the context of interviews, the ways people talked to me and to others, what they did with different texts or artifacts, and how they talked about them became an important source of cues for understanding what was going on in Trombetas with regard to literacy. These possibilities of contrasting indexical signs of changes related to literacy represent distinct ways of (re)entering the field and living the tension between *emic* and *etic* perspectives (Heath and Street, 2008) in the process of making sense of *what is going on* in Trombetas.

In the following sections, I will present initial results of the process of exploring these ways of contrasting indexical signs to present an account of changes related to literacy in Trombetas and examine their relationship with other contexts and social, economic, and historical factors. Finally, the consequences of these changes for how people have access to writing and the meanings of writing to them will be briefly discussed, drawing upon case studies on writing at work, and on the use of school literacy practices at home.

Global and Local Relationships: Situating Local Changes within a Broader Context

The first members of these families arrived in this neighborhood in the 1970s coming from rural areas. At that time, internal migration was intensifying as a result of structural changes in the Brazilian economy that was moving from an agrarian model to an industrialized one. In the late 1960's, new laws redefining rural labor contracts made working and living conditions even more difficult in the rural areas. In contrast, the new economic model aiming to promote the development of industrial areas in major Brazilian cities attracted families searching for new job opportunities encouraging inward immigration (Ribeiro, 2007; Scalon, 2004). The majority of migrants went to live in *favelas* or on the outskirts of big cities, where the infrastructure for transport, electricity, water, and sewers was minimal or non-existent. As an unskilled workforce and with little or no years of schooling, they were hired for very low pay, receiving a minimal wage.

Figure 7.1, from 1988, shows some characteristics of Trombetas 18 years after the first arrivals of these families in the area in the 1970s. At this time, the late 1980s, Brazil was going through economic recession: Trombetas continued without basic infrastructure and many people had lost their jobs.

In the 1990s the Brazilian economy gradually became more stable and inflation was brought under control. The industrial area of Belo Horizonte where Trombetas is located had confirmed the adopted economic policies with respect to industrialization and Trombetas became a dormitory neighborhood with many of its residents leaving daily to work in factories, commercial business, public services, or residences situated in faraway neighborhoods. In the local area there were also jobs in small business such as a mechanical garage, drugstores, supermarkets, and beauty salons.

In 2009 and 2010, changes in Trombetas' landscape reflect not only the changes in the urban infrastructure in this area but also the increasing economic growth experienced by the Brazilian economy in the last 10 years. Families have improved their houses or built new units on their lots for their adult children and their families. People interviewed expressed their satisfaction in experiencing some degree of employment stability, and many of them are no longer in the informal job market.

Analysis of schooling across the three generations of Trombetas families shows that the successive generations spent longer periods in school, partially, but not entirely, as a result of extension of obligatory schooling over this period. Most of the members from the first generation of these families, coming from rural areas, had very limited opportunities to go to school, and those that did go to school did not finish the primary level.

Among the second generation, schooling attainment was also very low: 50 percent of children failed in their first year of school, and many in successive years, so the situation of children from Trombetas was not different from what was going on in other parts of the country. Many of them, pressured by their "failure" in school and the need to contribute to their family's economic survival, were forced to leave school early, some in the early primary years, and others when teenagers, often joining the informal job market. However, among eleven children who participated in the first study and who later were interviewed by Lana (2008), five women had completed high school, and two had not finished elementary school; and five men had finished high school and two of them managed to go to a private accountancy college. Others that I have met, after leaving school to work when teenagers, went back to adult education programs to complete high school. Some succeeded, others did not.

Teenagers or children from the third generation of families interviewed in 2009 did not experience the same school force-out phenomenon as these previous generations did and did not seem to be pressured to go to work yet. The combination of educational and economic policies seems to contribute to the possibility of extended time at school. One of these policies is the adoption of automatic promotion in public schools[4] that corrected the flux of children in schools. Another is the *Bolsa Família* Program, initiated 11 years ago, through which poor families with children receive an average of R$70 (about U.S.$35) in direct transfers, and in return they commit to keeping their children in school and taking them for regular health checks. Children from the families interviewed are enrolled in this program.

Availability of Writing in Trombetas: Changes in the Landscape and Houses

The two pictures taken twenty years apart (figures 7.1 and 7.2) make visible some of the changes inscribed in the local landscape. While in 1988 the area was sparsely occupied, in 2008 the number and size of houses increased: some have more than one floor, and many families have constructed more than one house in their lot to provide housing for their adult children and their families. The roads that once were irregular and uneven are now paved streets. In the first picture, the hills in the back are still unbuilt. The broader view of the second picture shows how much the city has stretched its boundaries over the last two decades.

Kalman (2003; see also this volume) in her studies on participation and appropriation of literacy practices proposes the distinction between *availability of literacy* and *access to literacy*. *Availability of literacy* refers to the physical presence of printed materials and infrastructure for its distribution (e.g. libraries, newsstands, mail services, and so on). *Access to literacy* refers to opportunities for participating in literacy events as well as for learning how to read and write. This distinction will be explored in examining how changes in availability and accessibility may relate to possible meanings of writing among research participants.

Contrast between other pictures from the 1988 and 2009 archives allows us to see changes in the availability of writing (Kalman, 2003) in the streets of Trombetas.

FIG. 7.1 Trombetas landscape 1988.

Pictures from the school site in 1989 show that it was a prominent building on the top of a hill, surrounded by green hills. Currently, it is neighbored by commercial and residential buildings. There are many written signs hanging on these buildings, advertising products and services: material for construction, a 24-hour bank, and drugstore, among others. Someone approaching the school would see a tall wall where previously a fence marked the limits between the school and the outside world. The construction of high walls around the school buildings was an administrative answer to vandalism or robbers looking for schools' goods. It also served to control the students: before, it was easier to leave the grounds without being noticed. The wall, then, can be taken to index issues of security and control in the neighborhood and how they have changed.

In the late 1980s, writing was not yet very visible in the public spaces of Trombetas: for example, street signs or the numbers of the houses were scarcely present. Nowadays, streets' names are written on plaques and the houses are sequentially numbered in streets no longer interrupted by gaps. A news-stand placed in the main square close to the school gate is also an addition that occurred in the last 12 years. Writing has become available in newspapers and magazines for all ages and diverse interests. The stand's owner explained the profile of readers by age and social positions and, based on his knowledge, he provided a variety of the

FIG. 7.2 Trombetas landscape 2008.

cheapest publications that can address people's interests. The news-stand has a section of used paperbacks for renting or exchanging, attuned to the interests of women who, according to him, are mostly interested in love stories. In the 1980s the only library in the area was the school library, which was not in use even for the students (Castanheira, 1991).

The availability of writing in the private space of families' houses has also increased in number and diversity. In Table 7.1, I list the writing materials observed while interviewing research participants in their houses in 1988/1989 and in 2009.

The increased number and diversification of writing materials in the houses are also indexical signs of changes related to economy, new technologies, the presence of neo-Pentecostal churches, and new educational policies. For example, while there were no churches in Trombetas and most of its residents were Catholics in the late 1980s, by the 2009 visit I saw numerous neo-Pentecostal places of worship. These transformations were also marked by changes in the available artifacts inside homes. In 1990 I noted the presence of a single large volume of the Bible in a couple of houses. Nowadays, associated with the neo-Pentecostal churches in the area, I saw multiple samples of the Bible in the houses, presented in smaller sizes, with colored and designed covers addressed to boys, girls, or adults.

In the late 1980s, school materials such as old notebooks were used and reused many times by various family members, and textbooks were scarcely seen, and,

TABLE 7.1 Availability of literacy in the houses across time

Materials in the houses	
1988/1989	*2009/2010*
One Bible for household	Multiple Bible samples
Dictionary	Religious journals
School notebooks	CD/ DVDs
Re-utilized old school notebooks	Multiple textbooks
Small blackboard	Various notebooks (individual use; school & religious purpose)
Notepads for registering purchases in store	Small blackboard
Papers with children's drawings	Calendars
Calendars	Food cans
Food cans	Bills
Bills	Personal Letters
Personal letters	TV
TV	Radio
Radio	Recipe booklets
	Beauty products booklets
	Driver's license booklet
	Paperback books
	Frames hanging on the walls
	Personal cards on the walls
	Birthday and Valentine cards
	Newspaper
	Writing in construction site
	DVD & CD Players
	Video tapes
	Music CDs
	Computer

when they were present, they would be old and donated from the houses where some mothers worked as cleaners or maids. Nowadays, school materials have also multiplied; for example, children in all school grade levels have their individual textbooks for the various curriculum subjects (e.g. Math, Portuguese, History, Geography, English, and Natural Sciences). The presence of multiple textbooks in the houses results from changes implemented from 1996 in the National Textbook Evaluation and Distribution Program developed by the Brazilian Ministry of Education. In addition, the number of school notebooks also increased: students have different notebooks for different disciplines, and for Religious Studies. While textbooks are provided by the government, notebooks and other school materials (e.g. backpacks, pencils, or colored pencils) are acquired by families. The presence of DVD and CD players, and computers is also a sign of how the families are able to afford not only notebooks but also new technologies for accessing, consuming, and enjoying cultural products. A comprehensive discussion of how all these changes impacted *access to literacy* practices (Kalman, 2003) among Trombetas residents is out of the scope of this chapter. However, I will give a brief overview of changes concerning how family members now have opportunities for participating in literacy events, and present two examples to illustrate the relationship between writing and work, and the use of school literacy practices at home.

Writing, Reading, and Work

In the late 1980s, reading and writing were necessary to enable adults to read bills, take notes of debt in the market, receive and send personal letters, and identify which buses to take, for example. As many of them had little or no schooling, they found work as cleaners, masons, and street sellers with no formal contracts. According to adults interviewed in 1989, their jobs did not require writing. When talking about their children, parents expressed their concerns about not having money to give them a better life, including buying books or sending them to better schools.

Whilst second-generation men and women of these families continue to work as masons, house cleaners, or maids, some of them are working as messengers, teachers, or attendants. In their interviews, they stated that their jobs do not involve much writing, and that after a long journey to and from work they do not have energy for reading at home. Carla, who was interviewed when she was 9 years old, was interviewed again at age 29 (we will talk about her and her brother again later). Her first job was at 16, when she decided not to finish high school. By the time of the second interview, her son was 6 years old and she was working in a factory in a clerical job, an eight-hour shift, and on some days she stayed overtime. In an interview, she talked about her reading practices:

Interview with Carla on reading and work

Carla	*Researcher*
I remember when I was in the first year of high school	
I missed class in order to go to the library	
We could get 3 books at a time	
I would go with 3 friends	
each of us would get three	
	to exchange among yourselves?
No	
I would get them all	
After a while there were no new books for me to read in the school library	
I used to love books	
Did you know?	Ah
Now	
I fall asleep	

Carla was longing for the time she could enjoy reading lots of books again. Now, she feels too tired to read after work and taking care of things at home. Later in the interview, she explained that she "lost it/I have to read some parts over and over again to understand/before, I just read once." In her job, she does not need to read or write much: she just fills out forms on the computer. Her brother, Carlos, also started working at an early age, but managed to finish high school and attended a private accounting college at night. He was 27 years old, married, and had a 3-year-old daughter at the time of an interview in 2009. According to him, work took up too much of his time, leaving little time to read, or to watch a movie on TV, or rented DVDs. When his mother commented that he writes poems, he replied that he did not have time for that anymore either.

The rhetoric about literacy emphasizes its importance in the new economy, but economic changes in the region and the creation of a new job market did not really require special qualifications or higher levels of schooling from the workers interviewed. Many of the participants in the previous research did not conclude elementary school or high school and entered the job market very early. In this new scenario, writing is still seen as not required, or as being very limited in the workplace. Life and work conditions seem not to favor engagement with writing and reading activities related to other domains (e.g. leisure), and some may think they are losing the ability to read.

The Use of School Literacy Practices at Home

In 1989, writing was also present when parents and older siblings engaged in practices that they saw as preparing the younger ones to enter school. In the late 1980s, in many houses, traces of these practices were found in old notebooks reused many times to copy, for example, the vowels (a-e-i-o-u), the full alphabet, numbers, or words. Children were often seen, as they called it, "playing school," with one child taking on the teacher's role, while others (brothers, sisters, or friends) performed students' roles.

Analysis of old notebooks and playing among children made adults' and children's school experiences visible and served as a reference for choosing what to exercise as well as for defining what counts as "playing school." The contrast between materials used by Carla, a 9-year-old third-grade girl, in 1988/9, and by Clarice in 2009, when playing with their brother and friends, makes visible indexical signs of change in the approach used to teaching writing in schools over the last decades.

Carla usually played school with her brother when her parents were working, and some days other children would join them. Carla's parents had provided a small blackboard on which she wrote the activities to be copied and developed by her 6-year-old brother, Carlos, who was entering the first grade of elementary school, and their friends. She copied from a notebook on to the board the following sequence of activities (translated from Portuguese to English):

1) *Read these sentences:*
   Bia's father is very good.
   The stick is big and hard.
   The father is mad.
   Bia's dress is made of a very pretty cloth.
   The sink was placed in the kitchen
2) *Copy the alphabet twice:*
   A B C D E F G H I J K L M N O P Q R S T U V X Z
   a b c d e f g h i j k l m n o p q r s t u v x z
3) *Write your name once.*
4) *Study:*
   Pano, Pai, Papai, Pau, Pia
   AI, AU, EI, EU, UI,
   Eia, Uá, Uai, Ué, Oa, Oi, Ioiô, Iaiá.

This sequence of activities exemplifies the kind of teaching approach Carla and her friends explored in their playing. The "students" were expected to copy what she wrote on the board, even if the instruction was just *Read*, as in activity number 1 above. Her brother complained about this, saying that his sister gave them lots of things to do and, some days, insisted in continuing playing school even if he wanted to play on the streets with his friends.

Activity 4, for instance, resembles a methodological procedure present in early literacy textbooks used in the Brazilian schools in this period: a selected group of words from the "text" followed by a group of syllables are presented to exercise the recognition and memorization of relationships between phonemes and graphemes. The way Carla orchestrated the play time and organized and proposed activities for her brother and friends can be taken as evidence that she was re-enacting with them the ways she had been taught at school: copying in cursive letters and focusing on developing the ability to decode and write.

In 2009, I encountered a group of children playing school in a snack bar on a Saturday afternoon. Clarice, also a 9-year-old girl in the third grade, took up the role of teacher, while her 7-year-old brother and a 6-year-old friend (José) were her students. The snack bar owner was her uncle, and it was situated in front of her grandparents' house, which I had just left after interviewing them. A couple of tables were occupied by adults who would take turns going to the jukebox and selecting music.

Clarice's aunt works for the county Educational Department supervising kindergarten teachers, and she collects used school materials to bring home for her niece to play with with her friends. It was from her aunt's files that Clarice selected the photocopied pages of activities she gave to her brother and her friend that day. José was very engaged in the activities Clarice selected for him and would come back to the table she was sitting at to show her what he had done.

I had the opportunity to talk to Clarice and asked her to show me the activities she would choose for José from her aunt's file and explain why she would choose them. In one of the pages that she selected, a drawing represented a scene of a birthday party: cake (*bolo*), candies (*balas*), and balloons (*balões*) were placed on a table, surrounded by two girls (one of them blowing the candle) and a boy. This drawing was accompanied by the following instruction: *This is Bianca's party. It was a very fun party, like ours. Mark with an X what starts with the letter B.* Finally, the last activity proposed: "*Agora, escreva o nome de tudo que você marcou como souber. [Translation: Now, write the name of the things you marked in the way you know.]* The contrast between materials available to Clarice in 2009 and references explored by Carla in 1988 showed that two main methodological procedures are no longer present: sequences of phrases or decodable texts, and copying. Finally, a key innovation seems to be the proposition to the children to write something *in the way they know.*

These differences can be interpreted as an indexical sign of changes in methodological approaches to teaching writing adopted in Brazilian schools. The research of Ferreiro and Teberosky (1985), for instance, showed that learners create hypotheses about the uses and functions of writing as well as about the structuring principles of the alphabetical writing system. This finding led to the possibility of learners inquiring about, for example, his or her ways of conceiving how to write a word. This kind of question was inconceivable in the approach that was adopted in Trombeta's public school in the 1980s and that appeared in Carla's re-enactment of school.

Another indexical sign of change in the materials explored by Clarice is the presence of the word *letramento* (literacy) on the top of some of the pages in her aunt's file. This Brazilian Portuguese word was first used in an academic text in the late 1980s and, gradually, its usage gained attention in other social spaces (e.g. public school, media) (Soares, 2004). Its creation was a response to the growing recognition that the dichotomist alternative pair of words *alfabetizado* and *analfabeto* that referred to the ability or inability to read or write did not allow the possibility to talk about reading and writing as social practice (see Marinho, this volume). Thus, the presence of this new teaching procedure and the word *letramento* (literacy) on the top of the page selected by Clarice can be seen as indexical signs of changes in the approach to how writing is conceived and taught in Brazilian schools; for instance the attempt to take into consideration social uses of writing in the school teaching processes (Street, 2005). More research is needed on how these conceptual changes are taking shape in schools and their consequences for what and how one can learn literacy in and out of schools.

Conclusions

These brief accounts, based on observations and interviews, provide insights into how *the wider context* affected changes in the local context and what goes on in the life of Trombetas' residents in relation to writing and reading. The accounts can be read as signs of changes in the availability of and access to writing that took place locally in Trombetas,[5] in the city of Ibirité. I have called upon the concept of indexical signs as a way of analyzing intergenerational changes in both local and global terms, looking at signs of change in public spaces, such as the layout and design of the housing and roads and the place of a school within this neighborhood; the increase in written signs in these new spaces; and also the number of neo-Pentecostal churches, which has implications for the kind of religious texts available. I have used Kalman's notion of availability and access to consider the significance of such changes for local people. Finally, I have presented brief case studies of children bringing home school practices across the two periods of my field research – 1990s and 2009 – and related these to changes in the national educational system and curriculum. This chapter has mainly attempted to open up questions for further research in such contexts and to suggest ways in which we can relate local and global aspects of social change in general and changes in the uses and meanings of literacy in particular. Like many of the chapters in this volume, it looks forward to further studies of Latin American Literacy Studies that extend and build upon the current rich vein of research in this field.

Notes

1 The names of the neighborhood and people are pseudonyms.
2 This first case study was supported by Capes (1988–1991). The second received support from CNPQ and (Process 201834/2007–3), and FAPEMIG (159–09).
3 Many thanks to Priscila Maria de Lana, who helped me to get in touch with these families again.
4 Public schools in Brazil are those financed by the federal, state, or counties' financial resources.

References

Barton, D., Hamilton, M., and Ivanic, R. (eds.) (2000). *Situated Literacies: Reading and Writing in Context*. London: Routledge.

Baynham, M., and Prinsloo, M. (eds.) (2008). *The Future of Literacy Studies*. London: Palgrave Macmillan.

Bloome, D. (2005). Introduction to the study of classroom language and literacy. In R. Beach, J. Green, M. Kamil, and T. Shanahan (eds.), *Multidisciplinary Perspectives on Literacy Research*, (2nd ed.). Cresskill, NJ: Hampton Press, 275–92.

Brandt, D., and Clinton, K. (2002). Limits of the local: expanding perspectives on literacy as a social practice. *Journal of Literacy Research* 34, 337–56.

Castanheira, M. L. (1991). *Entrada na escola, saída da escrita* [*School Entrance, Writing Exit*]. Unpublished thesis, Belo Horizonte, UFMG: School of Education.

Castanheira, M. L., Crawford, T., Dixon, C. N., and Green, J. L. (2001). Interactional ethnography: an approach to studying the social construction of literate practices. *Linguistics and Education* 11, 353–400.

Ferreiro, E., and Teberosky, A. (1985). *Psicogênese da Língua Escrita* [*Literacy before Schooling*], Diana Myriam Lichtenstein, Liana Di Marco, Mário Corso, trans. to Portuguese from Spanish. PortoAlegre: Artes Médicas.

Green, J., and Bloome, D. (1998). Ethnography and ethnographers of and in education: a situated perspective. In J. Flood, S. B. Heath, and D. Lapp (eds.), *A Handbook of Research on Teaching Literacy through the Communicative and Visual Arts*. New York: Simon and Schuster Macmillan, 181–202.

Green, J. L., Dixon, C. N., and Zaharlick, A. (2002). Ethnography as a logic of inquiry. In J. Flood, D. Lapp, J. Squire, and J. Jensen (eds.), *Research in the Teaching of the English Language Arts*. Mahwah, NJ: Lawrence Erlbaum Associates (LEA), 201–24.

Gumperz, J. J. (1982). *Discourse Strategies*. Cambridge: Cambridge University Press.

Heath, S., and Street, B. (2008). *Ethnography: Approaches to Language and Literacy Research*. New York: College Teachers Press.

Kalman, J. (2003). El acesso a la cultura escrita: la participación social y la apropriación de conocimientos em eventos cotidianos de lectura e escritura [Access to literacy: social participation and the appropriation of knowledge in daily events of reading and writing]. *Revista Mexicana de Investigación Educativa* [Mexican Journal of Educational Research] 17, 37–66.

Lana, P. M. (2008). *Práticas de letramento na perspectiva de jovens adultos de camadas populares* [*Literacy Practices in the Perspective of Young Adults of Popular Origin*]. Unpublished thesis, Belo Horizonte, UFMG: School of Education.

Pahl, K., and Rowsell, J. (2010). *Artifactual Literacies: Every Object Tells a Story*. New York: Teachers College, Columbia University.

Reder, S., and Davila, E. (2005). Context and literacy practices. *Annual Review of Applied Linguistics* 25, 170–87.

Ribeiro, C. (2007). *Estrutura de classe e mobilidade social no Brasil* [*Class Structure and Social Mobility in Brazil*]. Bauru, SP: Edusc.

Scalon, C. (ed.) (2004). *Imagens da desigualdade* [*Images of Inequality*]. Belo Horizonte, Brazil: Ed. da UFMG.

Silverstein, M. (2006). Old wine, new ethnographic lexicography. *Annual Review of Anthropology* 35, 481–96.

Soares, M. B. (2004). Alfabetização e letramento: as muitas facetas [Literacy and *letramento*: multiple facets]. *Revista Brasileira de Educação* [Brazilian Journal of Education] 25, 5–7.

Street, B. (1984). *Literacy in Theory and Practice*. Cambridge: Cambridge University Press.

Street, B. (2005). *Literacies across Educational Contexts*. Philadelphia: Caslon Press.

Wortham, S. (2001). *Narratives in Action: A Strategy for Research and Analysis*. New York: Teachers College Press.

8

SURVIVAL OF ORIGINAL KNOWLEDGE[1]

Irma Rosa Fuenlabrada and
María Fernanda Delprato

The research (Delprato and Fuenlabrada, in press) that is the main reference for this chapter explores the problems of numbering and calculating faced by unschooled or underschooled indigenous craftswomen during their social and productive integration in spaces that are foreign to their culture. Understanding the underlying mathematical logic to these problems demanded venturing into an analysis of the social concerns that support the resolution strategies to the arithmetical problems that these women confront. The choice of research subject[2]—a collective self-managed productive organization—was linked to our interest in approaching the new mathematical tools that we supposed were demanded by the processes unleashed in the constitution of a productive organization that transcends the family domain. In turn, the singularity of the social endorsement of this context (a local context of indigenous communities) demanded the overlapping of these processes with community practices and cultural wisdom. This required discussing conceptions of "mathematical literacy" that are restricted to the autonomous acquisition of problem-solving skills. In other words, it entailed the adoption of an ideological model of numeracy[3] (Street, 1984) to analyze the link between the numeracy practices observed and the specificity of the researched context: that of a collectively managed indigenous organization dedicated to the production of handcrafted objects. In the following paragraphs we will recount the trading ways of the community (bartering and incorporation of the monetary system) and of the handcrafted production. We will also devote some space to the circumstances and events surrounding the creation of the organization of women artisans. In this account we seek to recognize new practices of numeracy and the power relations that cross them within the local context (the relationship between the leader and the artisans) and within the out of community social networks (for the defense of the value of handcrafted production). We will see that the tensions that cross and crossed this domain demanded the differentiation of preceding trading practices as a way to provide legitimacy to local power relations—what Elisa refers to as "the beautiful power."

The research was carried out by interviewing the leader of the organization and by accompanying her to a fair that took place in a town located near her community of origin (Pátzcuaro, in the state of Michoacán, Mexico) during the selling of the products handcrafted by her organization. This allowed us, from a perspective that was full of interpretative intentionality, to document the goals and meanings of the work of this woman leader. She was selected as informant in view of the peculiar position of intermediation that she held. We approached the interpretation of Elisa's goals reconstructing the constitutive link between the appropriation of the numeracy practices that are described and the identity processes of Elisa as leader.

This means that our analysis in the final sections will be centered on Elisa, on the construction of her position within the organization, and on the tensions between the numeracy practices evident in the exercise of her leadership, including tensions between the access, the interplay, and the resistance to new numeracy practices (and knowledge) within trading dynamics and out of the community trading spaces.[4] Throughout this process, we will be analyzing the acquisitions of skills and the cultural resistances that are part of Elisa's socialization process as the leader of the organization, and her ways of experiencing and valuing the exercise of this position. In short, we will be analyzing the constitutive elements of her identity[5] (Zavala, 2009). And since it is in the interest of this article that we focus the analysis on the link between numeracy practices and Elisa's identity, we will not go into more depth regarding mathematical resolution strategies.

Trading Relations in the P'urhépecha Community: The Original Trading Relations of a P'urhépecha Community and Trade Relations Mediated by the Monetary System

In Mexico, bartering for basic merchandise and subsistence foodstuffs had its origin in the economic relations that were established between social indigenous groups during the prehispanic era. Although this local trading practice is useful for those who participate in it, it does not enable access to the means of consumption in external economies regulated by the use of the monetary system. Bartering is still practiced in the organization that concerns us, but it now appears as a last resource for "selling" pottery, and not as the first and only one as it was in the past.

Currently, trade relations entail the use of bank notes (or peso bills in this case) and coins imply interactions with symbols and inscriptions for those involved. For the communities that are mainly non-literate, this has several consequences. First, if they do not interpret the symbols and the inscriptions on the bills and coins correctly, they put their vital interests at risk. Second, doing this obliges them to use mathematical tools to solve the differences of equivalence that regulate the monetary system, and they work this out from their own abilities in handling the numerical system.

Of the women in the organization, Elisa is apparently the only one who recognizes the equivalence relationships of the Mexican monetary system well; but this does not mean that she manages them with ease in trade transactions. Operating within the monetary system also requires the members of the community to understand a series of trading rules and mechanisms that are foreign to them. Given the above, the community members need to introduce the use of money within a complex process of specific adaptations, cultural adjustments, and communal decisions predetermined and redefined from within their own cultural perspective and local knowledge about trade relations.

Building a Collective in Response to Underprivileged Situations

Most of the women devote themselves to the hand production of *comales*[6] (clay griddles) and clay pots for daily use. The community also produces hand-embroidered *mandiles*[7] (aprons). The integration into the external economy favors the extension of pottery production to the making of "decorative clay objects"; in the same way, the embroidery of aprons is extended to the production of blouses.

In order to improve their family income, the women in the community need to establish trade relations with "the others," but while doing this they confront a double set of problems: (1) they have to acquire a new language—Spanish—since they are P'urhépecha speakers; and (2) it is difficult for many of them to attend markets to sell their handcrafts because many of them cannot leave the community. Both situations restrict the participation of several of the indigenous women in the selling process.

In the past, these limitations led to disadvantageous negotiations between the craftswomen and intermediaries who were foreign to their community and who took over the outside sales[8] of their products. In a previous study about the group, Torres (1995, 43) points out that: "Before the constitution of the group, sales were local and through some intermediaries, whose management and control did not allow for the improvement of their conditions and demanded very low prices for the products of the women potters."

The search for more just trading mechanisms for their handcraft production propelled this group of women, under Elisa's initiative, to organize themselves and form a collective. Initially, the craftswomen of this community were only looking for a fair price for their products through the intervention of a community member who could play the role of sales agent towards the outside markets. That role fell specifically on Elisa. It is important to note that neither Elisa, nor her companions (Mananei and Araceli), received any payment for this endeavor. The "intermediary" role that they accepted stemmed from their sense of solidarity towards the other craftswomen in the community.

The Commercialization of the Pottery

Beside other problems, the "intermediary role" in the selling of pottery, adopted essentially by Elisa, revealed some difficulties for the women as time went by. It was enough for one group member to question the way in which her products had been traded to generate suspicions from the other members.

There is, in fact, a difference between the assumptions of the community members who do not go out to sell about the way that the sales are carried out and the actual experiences of those who are in charge of selling. To sell outside the community implies expenses, and the added risk of not being able to sell, or the possibility of lowering the price, and in some extreme cases resorting to bartering. This is difficult for some craftswomen to take into account if they only think in terms of the price that they would like to get for their products. In this context, money becomes an important part of the mechanism that mediates the exchanges between "the outside" and the community.

Taking this into consideration, Elisa and a few other artisans choose to buy the pottery from other members of the organization, paying them a fair price, although this represents a risk for them during the resale process (the risk of paying transportation costs from their pockets, selling at an equal or smaller price than the one they bought the piece for, or the risk of having to release the product through bartering).

Buying for reselling is essentially a way to bring transparency to the trading transactions from a position of justice. The craftswomen who can go out of the community to sell, buy and pay a fair price to the ones who cannot leave the community to work at the market. By doing so, they give their fellow members the opportunity to trade while safeguarding them from intermediaries who are foregin to the organization, and who pay a price for their goods that does not correspond to the value of ther wares.

In a different way from those who make pottery, the craftswomen who produce textiles and are able to leave the community do accept selling blouses and aprons through intermediaries. It is quite paradoxical that there are such differences in the selling processes within the same organization, especially because the saleswomen in charge of marketing the pottery and the textiles are the same. This situation is due to:

- the quality of the embroidery of the aprons (which are part of the traditional dress), which is the reason that the sale price remains high (reaching $2,000 pesos):[9] a price that prevents exchanging aprons for other merchandise in cases where they are not sold;
- the blouses could be sold more easily due to their price (at around $300 pesos), however, these blouses are at a disadvantage when compared to equivalent products of indigenous women from other regions, whose embroidered blouses are more original in design.[10]

The lack of competitiveness and the cost of production translate into smaller sales for the blouses and aprons. This does not justify the presence of the members of this textile subgroup to make direct sales because it is difficult to recover the expenses involved in traveling away from the community. In addition, *bartering* becomes impossible. This marks a clear difference with what happens to the pottery pieces, which have prices starting at $10 pesos ($1).

Elisa takes on the commitment of buying her co workers' pottery production and then reselling it with an attitude of solidarity with the craftswomen, especially because she recognizes that she has had opportunities that put her in a different situation from the rest of them. She therefore decides to take this into account by benefiting her fellow craftswomen with the opportunity to sell their products.

Elisa's authority ("the beautiful power") is founded on the "gift of giving to the other," a social criterion recognized in a variety of studies (Torres, 1995) as characteristic of indigenous communities. Underlying this criterion lies the value of reciprocity in attitudes related to giving and receiving, as well as in the fair distribution of goods. This not only refers to material objects, but first and foremost to those gains that come from personal experience and to the knowledge one has acquired through performing a different role. These non-material goods are not considered personal property or obtained by one's own merit; therefore they should be shared for the benefit of those who have not had similar opportunities.

The Rise of a Self-reliant Productive Organization

The fact that the community craftswomen organized themselves into a collective to facilitate the trading of their products without the involvement of external agents facilitated the transition from individual (or family) production to a coordinated group production.

The collective solution for managing the distribution of costs and profits is one of the main implications of this change. However, because the women aspire to consolidate their organization, the implementation of certain criteria for making sales-related calculations becomes more important. It becomes necessary to implement mechanisms agreed upon through collective consensus in order to guarantee a shared understanding of the new trading rules that are needed given the collective production of their wares; rules that are different from the ones that regulate individual production and that are best known to the organization's members.

When the members of the collective find themselves involved in procuring financial credit, an unavoidable differentiation arises among them. In Mexico, in order for a group of craftswomen to acquire credit, the group should be integrated within a government project. The regulations of these credit agencies establish the need for the existence of three positions within the organization: President, Treasurer, and Secretary. This implied the differentiation of the positions of the women within the organization, and gave way to what is formally known to the outside as a productive self-reliant organization.

A Woman Committed to her Culture of Origin

It is not enough that outside agents imagine *a priori* the establishment of a productive organization in an indigenous community for them to be able to provide this organization with hypothetical economic benefits. And the fact that the leader of the organization belongs to the community will not guarantee that it will function properly either. One condition—among others—that might offer the possibility to create a productive organization capable of improving community living standards is the participation of a person with leadership qualities, whose personal trajectory and projections contribute to the organization's development.

The organization, as has been pointed out, finds in its leader, Elisa, a well-known and respected woman within her social environment. Elisa establishes the organization of women potters and textile producers in her community and, from the beginning to date, acts as its president. The positioning of Elisa as leader is linked to what she learned within her family nucleus. She is the daughter of a peasant who, accompanied by other indigenous people, negotiated the acquisition of fertilizers from government agencies.

Her father's leadership within the community is part of Elisa's family legacy. At the same time, it seems that she is not subjected to the submissive character of women in her social milieu. She is 46 years old and single, and does not seem constrained by this situation. This has allowed her to move about socially and to make decisions that are closer to what masculine community positions might be, rather than to those of her own gender. Elisa assumes leadership among those who share her gender's condition, without upsetting the inherent circumstances of the women she leads and organizes in their own economically productive activity. In addition she holds strong solidarity ties with her culture of origin.

As Spanish is her second language, she experiences certain limitations for communicating with others and defending herself as an indigenous P'urhépecha. At the same time, these difficulties coexist with the pride of being who she is, and the benefit of producing handcrafts different from those traditionally made in the community, which was a vision she had when she organized her playmates at the time when she was still a child. Elisa's childhood experiences created for her a distinct position within her family. This, in turn, crystallized as leadership outside the family with demands that distinguish her from other women in her community who are either non-literate or underschooled.

Surviving the Knowledge of "The Other"

Elisa's leadership, then, positions her as the intermediary between "the outside" and "the inside" of the organization and from that position she learns to understand the knowledge of "the other." In this way she comes to terms, in a gesture of recognition of her traditional knowledge, with the appreciation of traditional productive and communal organizational practices (see Knijnik, this volume).

We will now proceed to briefly analyze how the role of President compels Elisa to acquire new knowledge in order to carry out the duties of her position. Similarly, we examine how she comes to recognize that the knowledge of the others—Spanish language, accounting, reading, and writing—has social value. At the same time that she becomes aware of her limitations and her ability to access these types of knowledge, she finds specific ways to understand and deal with them.

Elisa had her first trading experiences accompanying her grandmother throughout the rural settlements that were near her community. In her journeys outside the community, her grandmother was attacked verbally by Spanish-speaking neighbors for her deficiencies in her ability with the language, and from those experiences Elisa began to recognize the "limitations" of her mother tongue as a means of communication for interacting with "the outside."

For Elisa, learning Spanish was initially a tool for making herself understood and getting along with "the others" while selling pottery outside of her community. While other indigenous people have felt ashamed of their language, Elisa responds to and confronts aggressions with pride in being a P'urhépecha speaker. She realizes her identity in her language, and, moreover, she enriches her mother tongue by incorporating expressions used by other P'urhépecha-speaking communities.

Apart from preserving her mother tongue, Elisa is interested in preserving the original manufacturing practices for pottery-making in her community (kneading the clay and manipulating it with bare hands until the piece attains the desired shape) despite the disadvantages of the physical damage caused by these methods, and also despite the fact that, more and more frequently, buyers give little value to "the way" clay pieces are made. They are reluctant to absorb the impact of the traditional way pottery is made on the price they pay for the clay pieces.

Distribution of Opportunities

Elisa is responsible for negotiating the spaces for the handcrafts' sale at the fairs that take place in towns with bigger local and/or touristic affluence (Pátzcuaro, Uruapan). Also, attending these fairs presupposes the organizing of the cooperative's members for their participation in the contests organized in these spaces by the Casa de las Artesanías del Estado (the House of Handcrafts for the State of Michoacán).

Winning a contest represents receiving significant cash income for all involved. By community agreement, part of the prize is awarded to the artisan and part is given to the rest of the participants in the contest. Also, the winning piece can then be sold for a larger amount at exhibitions of participating handcrafts.

In order to assure fairness, Elisa manages the *sharing of opportunities* within the contests: only one member of each family can participate. Within this criterion lies an anticipation of an *egalitarian distribution* of the economic benefit expected to be obtained when winning the contest. Elisa maintains this criterion independently of family ties. This explains why her sister Romelia did not send a piece to the contest: *It is better that they decide, either the son sends or... the mother sends.*

Elisa also negotiates the selling of textiles and pottery from the community to governmental agencies, such as the Casa de las Artesanías de Morelia, en el Estado de Michoacán (House of Handcrafts of Morelia, in the state of Michoacán). In these negotiations, the control of merchandise requires reading and calculations in order to regulate sales. Elisa relies on the knowledge of external agents (buyers add up bills and write receipts for her); however, this does not mean that she disregards her own ways of being in control: *I also do the accounting ... at times I rely on my fingers.*

She knows that the one who sells *earns more* because with time she capitalizes on her experience in the selling process, but she also understands that for the craftswomen who have not had this opportunity, this can raise doubts and concerns, such as suspecting that the middle salesperson "is earning more" because "she keeps part of the sale for herself that she does not report to the producer."

Elisa takes on the buying and selling of pottery recognizing that she has a better chance to make sales derived from her experience in interacting with "the outside." This experience provides her with the capacity to identify the merchandise that is more saleable, to be innovative, to recognize and understand the sizes of pots more in demand, as well as to recognize the disadvantage of showing the price on the merchandise, because this hinders the *bargaining* process for the price of a product between the seller and the prospective buyer.

The resale of pottery puts Elisa in an atypical situation: she is the one in charge of the sale independently of the fact that the pieces wanted by the buyer belong to other women that are at the stand. During the negotiating process there may occur some bargaining for the purchase of one or even several pieces; Elisa then asks for the owner's or owners' consent; then she works out the sale's total; and, in order to avoid complicating the calculations afterwards, she immediately distributes the corresponding amounts for the pieces sold amongst the craftswomen involved.

Redefining the Functions of the Organization's Representatives

We have mentioned that the credit agencies demand that these types of organizations appoint three representatives. In the present case these were President (Elisa), Treasurer (Mananei), and Secretary (a young student graduate from the secondary school in the community).

The leader and the other two individuals, with whom she shares the organization's responsibility, use organizational schemes that are foreign to them and imposed from outside. However, the tasks and competences of these representatives have been adapted—by the President—to the practices, customs, and opportunities that the social relations within the community can accept and recognize, the result being that they become different compared to the ones anticipated by "the outside."

The strong bonds of solidarity between Elisa and her people and the revaluation of her indigenous roots are ingredients for the shaping of a leader who is not *ladino* (of mixed race, Spanish speaking) and who is capable of exercising power based on fairness. This underlies her notion of leadership and has made it possible for the President of this organization to organize her community with a vision oriented towards improving the quality of life within it. She does this by means of introducing and adapting cultural assets from "the outside," from a clear position of aspiring to preserve the cultural assets of the community. This is how Elisa makes decisions: based on the consensus of the members of her organization, she distributes the economic benefits of the commercialization from a perspective of equality of opportunities (she buys pottery from the artisans who cannot go out of the community to sell what they produce and takes charge of the sale of textiles without getting any profit for herself). She also distributes the opportunities to participate in handcrafts contests among the members of the organization, applying the same fairness criterion.

For Elisa, President of the organization, holding this position has demanded that she interprets and uses symbolic systems: numbers, mathematical operations, the writing of Spanish, and, to a lesser extent, reading in this language. The organization's presidency placed in Elisa the need to comprehend and establish an interrelation with the resources that "the outside" has incorporated into commercial transactions; amongst them, asking for credits stands out. This involves some complex questions that will be described next.

Collectively, the organization's members decide how much of a loan they are going to ask for; anticipating their ability to pay it off on time based on a participative distribution of the possible credit benefits. In other words, "every one has to get something—the same to all—interests included." Government Agency credits are granted for buying raw materials; this is why the agencies want to see the invoices in order to justify allotting credit to the organization. And because the craftswomen's participation in acquiring the credit is equal, consequently, so should the expenditures on raw materials be, but this doesn't happen as directly as one may suppose.

The buying is done collectively to bring down costs, and for every purchased raw material they get an invoice. Elisa is the one who requests and follows up on the invoices because she is the one who needs to present them to the agents who approve the credit. Once the credit has been granted, in order to maintain the organization's trustworthiness with the credit institutions, one of the President's tasks is that of collecting the money to pay the debt when the credit period ends.

The role of Mananei, "the treasurer," is that of a companion. Elisa recognizes this: ... *all the time with her. She goes with me everywhere.* Besides, she is in fact someone who certifies the President's word. This is not to say that her role is to supervise, but to respond to Elisa's need to give transparency to the different events that need to be reported, since they happen outside the community and directly affect the organization's members.

Another one of Mananei's roles is to assist the President by becoming "her memory." She performs this task during the selling process by keeping mental records of what goes on during it. (…) *when we sell something*, Elisa points out, *we go together.* (…) [and I tell her:] *look, we are selling this for so much.* (…) *She is aware that we are selling* [a product] *from a person, of how much it costs, how many blouses we are selling from a person.* Then, when Elisa needs to hand over to the textile craftswomen the money from the sale of their merchandise, the Treasurer, with this memorized information, takes on the role of the one who corroborates the President's accountability: *So this is what she …, only being present…, and she also sees in* [her] *memory and tells me: yes, we sold it in so much and you were going to subtract so much from her. And like that, yes, like that we deliver* [the calculations].

Credit institutions do not foresee the need to help community organizations with the "paperwork" required to apply for credit. This is why Elisa defined the secretary's functions taking into consideration the calculation skills, as well as the reading and writing capacities, of the members of the organization.

> I also have one … secretary and since she did complete secondary school … she maintained complete control. When I told her: make a note, they will give us a loan on such and such a date, and on such and such date we will pay and this will correspond to the interest. And she would take notes and it was all written down.

In any commercial organization, numbers and calculations appear to be the mediators in the sales that are supported in the monetary system. In the next section, we will briefly emphasize the way in which numbers are used by these women, particularly by Elisa, in the arithmetical affairs that they will inevitably encounter.

Numbers and Calculations

In the situations that call for the use of mathematics within the organization, one can recognize the everyday knowledge about numbers and calculations that is realized there. Next, we will describe some of those social uses reconstructed through the ongoing support and participation of events in which there is the use of numbers and calculations linked to the production and commercialization achievements of this organization of handcraft producers.

The traditional community numerical series was replaced with the numerical series in Spanish when the community's business relations with the outside expanded, or as an effect of the community disintegration (because of the migratory phenomenon). Elisa explains that:

> in our language [the numerical oral series] I did learn it from my granny. I only know until one hundred [11] ... If you would ask a young boy, he would explain it to you in language, in Castilian [Spanish]. Here nobody knows how to count [in P'urhépecha] anymore.

There are three stages in the trading operations carried out by the organization: pre-sale (credit negotiations and price setting); sale (the resale of pottery, the sale through intermediaries of textile products, bargaining and bartering); and post-sale (distribution of prizes and of the value of the sales of blouses and aprons).

Here we only briefly analyze the ways in which Elisa, with her mathematical understanding, attempts to resolve these situations in accordance with her social criteria of transparency, fairness, and "giving to others" that rule her particular vision of the exercise of power derived from her position of leadership.

Written Numerals (Indo-Arabic)

Written numerals appear in pre-sale situations as well as at the time of sale. They serve as control mechanisms for loan payments' due dates and the corresponding interest belonging to each of the organization's members, when the President's memory may not be sufficient to remember. At the same time, occasionally they serve as a resource to solve conflicts between what a craftswoman may suppose she owes, and what the President may say about what she really owes, even if the document which contains the data cannot be read by one of the parties.

Written numerals are also a way to control the merchandise received and the sales performed. When Elisa receives merchandise, she asks the Secretary to make a written record, which is complemented by the use of tags that are attached to each item. The tags have the initials of each craftswoman's name, along with the price of the product. When a textile product is sold, the President attaches the corresponding tag on to a piece of cardboard and in that way she can reconstruct what happened when she is back in the community.

Addition

Addition appears at the moment when product prices are fixed. These are linked to the task of assigning and setting a price to the value of the work invested in the manufacturing of the product. Prices are calculated according to what the craftswomen expect to earn by the day ... *it's about that more or less ... it's enough, it's not enough*. The capitalist system has solved this problem in a less complex way because, among other considerations, labor costs are established by a third party.

The setting of prices is one of the most complex tasks. However, within the organization, solutions are found by utilizing a complex set of compensatory adjustments that involve the experience of possible sales of different handcrafted products, and the need to "standardize" the prices (in multiples of five) to diminish the calculation requirements at the moment when the trading is taking place.

The calculation of the total of each sale is done by the President because she is the one who speaks Spanish. Right after receiving the money (especially for the sale of the pottery), she makes its distribution *in situ* (among the saleswomen who are at the stand). That saves her the need to write a sales' control record for distributing the money afterwards.

Division

The loan distribution is resolved "beforehand" because the craftswomen collectively decide this, based upon speculations regarding their ability to pay; and, after these considerations, they make adjustments in order for all of them to assume the same commitment and receive the same benefit.

Subtraction

Subtraction appears in bargaining situations and faces the President with the need to make decisions about possible discounts and, consequently, to make subtractions, even though the possibility of losing control over the profit causes her concern. Elisa can only operate with the number five or with multiples of five; she recognizes this limitation and tries to control it with approximations that can be observed in some of the decisions that she makes in bargaining situations. For example, she does not lower the price of jars that cost $10 pesos, as she feels that she would lose too much (50 percent); however, if the customer buys four jars or more, she reduces $5 pesos from every jar of $20 (losing 25 percent); or she sells a jar of $100 for $60 (with a 40 percent discount) because the buyer is an indigenous woman. If she sells jars of different prices to one buyer alone, the situation becomes quite complex, because she applies discounts of $5, $10, $15, etc. to each jar and not to the total of the sale.

The resource that would guarantee control over the profit would be to apply a proportional discount to the cost of each piece, which would be equivalent to applying this discount to the total of the sale; however, Elisa does not possess this knowledge. The lack of knowledge about proportional correlations becomes the main limitation regarding the mathematical solutions observed in the organization's trading transactions. Regardless of the fact that this skill is not available for the women to solve problems that occur within the organization—distribution of expenses, raw materials, prices—this is less serious than its absence in the negotiations with "the outside." In the internal exchanges, the resources for compensation and adjustments (alternatives to proportionality) are satisfactory because they respond with greater emphasis to the community social criteria, while in business relations with "the outside," particularly in bargaining situations—recurrent in these types of sales—the lacking of this skill, as has been explained, puts the craftswomen in a serious disadvantage.

In Conclusion

As we have observed, the incorporation of a collectively managed indigenous organization into a market sustained by the logic of accumulation and not by that of subsistence favors the emergence of conflicts and adjustments that are gradually regulated by solidarity ties sustained by the community practices forged throughout their social history.

From the preceding analysis of the tasks and roles of the three representatives that constitute the organization's leadership (President, Treasurer, Secretary), one can deduce that the President's responsibility entails unequal demands. Among these, the most significant are: a great negotiation capacity; a greater command of Spanish language; a broad use of her recall capacity; a reasonable understanding of numeric calculations; and a great capacity to argue and to communicate within as well as outside of the community.

This President exercises a type of leadership that manages the traditional knowledge and possibilities of the women who share with her the organization's responsibility in the leadership positions. She also manages the traditional knowledge and possibilities of the members of the collective in accordance with her vision of the organization and its needs. This is how she defines the companionship task of Mananei as treasurer, and the record keeping she delegates to the Secretary in view of her higher level of education. This is also how she regulates equal participation in sales in relation to the craftswomen' production capacities. In the same way, she also manages the benefits of participation in and the winning of "handcrafts" contests ahead of time, by "distributing" among the families of the organization's members the opportunity to participate.

We have endeavored to present a brief overview of the transformations of the social relations that take place within a group of craftswomen who decided to organize themselves for establishing commercial ties with an unfamiliar social environment that tends to impose on them trading methods that are quite different from their own. This is how the interpretations, adaptations, and adjustments that these P'urhépecha women have come to adopt are revealed, within the organization that unites them to interact with "the others" and "the outside." At the same time, it becomes clear how the cultural distinctiveness of the community serves as the support these women have for not surrendering themselves to the trading logics built from outside perspectives.

Notes

NOTE: The authors of this chapter have used pseudonyms throughout to protect the identities.

1 This chapter is based on a version presented at the annual meeting of the American Educational Researchers Association that took place from April 30 to May 4, 2010, Denver, Colorado, United States of America. Thematic group: *Social Contexts of Education*: Present Themes in Language, Written Culture and Mathematics in Latin America. Research from Mexico and Peru.

2 For a review and discussion of these distinctions in the field of numeracy, see Baker, Street, and Tomlin (2003), and Fonseca (2009). Starting from the conceptualization of numeracy as social practice, these authors recover some analytical consequences of this perspective that are outlined in the field of literacy. They also delimit some key concepts in the field of numeracy (Baker et al., 2003; Street, Pahl, and Rowsell, 2009): numeracy events and numeracy practices. This distinction between events and practices entails the inclusion of questions that exceed the mere observable behavior, such as the ones related to social relations of power. Consequently, in the contrast between school and everyday numeracy practices, one can find differences not only in the procedures, but in the unequal hierarchy of these practices, which result in the possibility of concealment of daily practices in the school domain.

3 This neologism comes from an analogy with the "*literacidad*" ["literacy"] that has been introduced in the field of New Literacy Studies in Latin America, to "fill a semantic emptiness in Castilian [Spanish]" creating a term that differs from that of "alphabetization" [basic literacy] because "… it is usually restricted to learning that is technical and out of context within the educational domain. On the other hand, the word is linked to negative terms (such as illiterate and illiteracy) that are strongly loaded with an official ideology that associates them with the lack of 'progress' at the social level and with the lack of 'intelligence' at the cognitive level" (Zavala, Niño-Murcia, and Ames [2004], 10).

4 The distinction between primary and secondary discourses made by Gee and reviewed by Kalman (2008) is very useful for this analysis, since "… the key point of secondary discourses is that, by definition, they involve the interaction with people that are out of our daily spheres of confidence, people with whom one cannot assume big quantities of shared knowledge and experience, or who involve 'formal' interactions; that is, people who hold an identity that transcends the family or the primary group of socialization" (Gee [1996], 143; mentioned by Kalman [2008],126).

5 It is necessary to call attention to the fact that we adhere to a conception of a non-essentialist identity. Therefore, the features reconstructed here must be "read" within the framework of the circumstances analyzed, simultaneously admitting the possibility of performing multiple identities in other circumstances (Niño-Murcia and Rothman [2008]. Quoted by Zavala [2009], 355).

6 The *comal* is an almost flat utensil of concave shape that is used for making corn tortillas. It is widely used throughout the rural zones of Mexico. They are also made of tin (metal) and these are also used for frying.

7 The *mandil* (apron) is a piece of clothing that is worn on top of the skirt (or of the dress); in this case it is decorative but its purpose is usually to prevent the skirt from getting dirty.

8 For the purposes of this document, the terms "external resale" and "external resellers" are used to designate the action of reselling and the individuals who are foreign to the organization and who buy the handcrafts for a very low price to sell them outside the community …

9 In 2005 prices (equivalent to U.S. $200).

10 This was observed at the Pátzcuaro Fair on the Day of the Dead. This festivity attracts national and international tourism. The "lack of originality" of the blouses is explained

by external social agents, who proposed venturing into this market; that is why there is no established tradition that would make them particularly attractive in the marketplace.

11 *"The first number is called 'ma,' 'tsiman'* (two), *' tanim'* (three), *'t'am'* (four), *'ium'* (five), *'kuimo'* (six), *'kuimo ma'* (seven), *'ium tanim'* (eight), *'ium t'am'* (nine), *'tèmbin'* (ten), *'tèmbin ma'* (eleven), *'tèmbin tsiman'* (twelve), *'tèmbin tanim'* (thirteen), *'tèmbin t'am'* (fourteen), *'tèmbin ium'* (fifteen). *And this way ... in fifteen and already first with that already and then the other ... Uhm ... hundred is ... this ... 'ma irépit.'"*

References

Baker, D., Street, B., and Tomlin, A. (2003). Mathematics as social: understanding relationships between home and school numeracy practices. *For the Learning of Mathematics* 23 (3), 11–15.

Delprato, M. F., and Fuenlabrada, I. (in press). *El poder de "las cuentas": poder con las cuentas y las cuentas del poder problemas de cálculo en la comercialización y preocupaciones sociales de una líder indígena.* [*The Power of Counting: Being Able to Count, and the Counting of Power—Problems of Calculation in the Trading Practices and Social Worries of an Indigenous Leader*]. Pátzcuaro, México: CREFAL.

Fonseca, C. (2009). Conceito(s) de numeramento e relações com o letramento [Concept(s) of numeracy and relationships with literacy], in C. Lopes and A. Nacarato (eds.), *Educação matemática, leitura e escrita: armadilhas, utopias e realidade* [*Mathematics Education, Reading and Writing: Armadillos, Utopias and Realities*]. Campinas: Mercado de Letras, 47–60.

Kalman, J. (2008). Dimensiones conceptuales en el campo de la cultura escrita [Conceptual dimensions in the field of written culture]. *Revista Iberoamericana de Educación* [Iberoamericana Journal of Education] 46, 107–34.

Street, B. (1984). *Literacy and Theory Practice.* Cambridge: Cambridge University Press.

Street, B., Pahl, K., and Rowsell, J. (2009). Multimodality and New Literacy Studies. In C. Jewitt (ed.), *The Routledge Handbook of Multimodal Analysis.* New York: Routledge, 91-200.

Torres, N. (1995). *Alfarería, organización de mujeres indígenas y aprendizajes* [Pottery, organization of indigenous women, and learning]. México: CREFAL-PEMT-OEA.

Zavala, V. (2009). "¿Quién está diciendo eso?" literacidad académica, identidad y poder en la educación superior ["Who is saying that?" Academic literacy, identity and power in higher education], in J. Kalman and B. Street (eds.), *Lectura, escritura y matemáticas como prácticas sociales: Diálogos con América Latina* [*Reading, Writing and Mathematics as Social Practices: Dialogues with Latin America*].Mexico City: Siglo XXI Editores and CREFAL, 348–63.

Zavala, V., Niño Murcia, M., and Ames, P. (eds.) (2004). *Escritura y Sociedad: nuevas perspectivas teóricas y etnográficas* [*Writing and Society: New Theoretical and Ethnographic Perspectives*]. Lima: Red para el Desarrollo de las Ciencias Sociales en el Perú [Network for the Development of Social Sciences in Peru].

PART III

Literacy and Numeracy in Education: Experiences in Latin America

9

WHEN LITERACY BRINGS TOO MANY RISKS: A SUCCESSFUL LESSON IN FAILURE

Shirley Brice Heath and Daniel Sobol

Literacy Programs and Social Mobility

Since the beginning of literacy programs designed to spread reading and writing beyond a small elite group, the matter of "benefits" that come with literacy has rarely been seriously questioned. Initially linked in Latin American nations with religious goals of friars and missionaries, reading and writing grew to be associated with social privilege and a rise in social status. Tightly associated with the dominant language of each nation, the written word in that language promised to open doors for work opportunities, land ownership, political power, and marriage possibilities. However, these openings have seldom materialized for the millions living in poverty with little chance of advanced schooling or social mobility in the legitimate economy. As researchers around the world have shown in their long-term statistical studies of literacy rates in correlation with growth in capital and opportunity, strong intervening political and social factors easily derail individuals who hope to move up the social class ladder. Some of these factors relate to language competencies; others have strong ties to regional and racial barriers. In the contemporary world, the hopes of the young are derailed when their formal schooling inadequately erases markers of class origin that will not move easily into the kinds of employment (generally in urban centers) that appeal to young people today. In this chapter we provide an example of such a case in which a community gardening and literacy development project in the Dominican Republic seemed to offer the opportunity for social mobility, but which in the end was overtaken by the kinds of social capital constrictions we have indicated here (and that are described in many of the other chapters in this book).

A Community Gardening and Literacy Development Project in the Dominican Republic

We could barely contain our excitement. In the summer of 2009, after months of planning and preparation, our project in community gardening and literacy development was ready, and our team of five young men was heading to the Dominican Republic (hereafter D.R.). As is the case with many recent college graduates who come from privileged backgrounds, these young men wanted to apply their education to benefit others. In the D.R., they would implement a pilot program to benefit the education and career chances of adolescents from working-poor families in an urban center. They would collect a treasure trove of data from a project we felt sure would prove to be meaningful, sustainable, and lasting in its impact on the D.R. teenagers.

We (Heath and Sobol) had written the design experiment in community gardening and literacy (Shavelson, Phillips, Towne, and Feuer, 2003). We had run training workshops with the four young men (P, N, Z, and A) who were to work with Sobol on the ground in the D.R. We had carefully planned the collection of data on the learning achieved by adolescents in the program. We had wrestled the learning sciences theories behind the project into practices that we believed would appeal to teenagers. Sobol had spent time over several years in the D.R. city in which we were launching the project, and authorities of the school these adolescents attended had helped Sobol craft the program to be of value to the school as a whole. We had studied the trappings, obstacles, and failures of development work focused on education in Third World countries. In designing our project, we had sought to minimize our ethnocentrism and top-down assumptions.

Drawing heavily on research in science education that emphasized the positive value of drawing and botanical illustration for students' engagement with science, the project centered on working with a group of teenagers from our partner school in the D.R., and creating with them a community vegetable garden within the school grounds. In association with the physical work of the garden, done in the cool morning hours, the students would spend the hottest part of the day learning botanical illustration and photography. Students were to learn to look closely at botanical illustrations of plants and insects from other parts of the world to help them think about details of the local plants they knew in order to illustrate these in the printed manual documenting the science of the garden for use by younger students. They could also use these noonday hours to learn film animation and storyboarding to create a short animated film about the development of the school's garden they were spearheading.

Students would be asked to think of a form in which they would present what they were learning. Suggestions we offered ranged from a book of photographs to a website to a short video directed by N, a graduate student in art and design, and A, a college graduate experienced in youth education. P and Z, specialists in horticulture, led the physical work needed to build and maintain the garden, and spearheaded the scientific learning sessions on biodiversity, plant sciences, and basic environmental science. Sobol coordinated the project on site to ensure coherence of skills, activities, and interests most likely to draw the students into ownership in a project that offered the teenagers opportunities to learn the following:

1) Web design
2) Photography and video production
3) Botanical illustration
4) Outdoor activities with peers within the safety of the school grounds
5) Production of food for the school cafeteria where their younger siblings ate each day
6) Design of the garden, its fencing, gate, and composting area
7) Recognition of their new roles of responsibility by school administrators, teachers, and guests
8) Travel to other sites in the city where gardening projects were underway through state and regional agricultural programs
9) Creation of a collaborative project of their choosing to reflect the garden project
10) Opportunities to work as science aides in classrooms of primary-level students.

During his visits to the school before the project's initiation, Sobol had learned that the school currently offered students none of these possibilities.

In the city of our project's location and in many other cities of the D.R., adolescents of working-poor D.R. families attend school only half a day, because of the limited space and staffing for the upper grades. We held the premise that if older students could be of use within the school during the half day when they were usually sent home, they could practice their literacy skills with younger children by drawing, reading, and helping teach them about environmental sciences. The working–learning linkage of older students with younger learners would help build a sense of community within the school and help young children feel a sense of ownership in the garden and cafeteria. They would be able to see their classroom learning all around them. The garden was to be a learning laboratory for younger and older students and their teachers, parents, and guests.

The overarching content focus of the project centered in the literacies, alphabetic and mathematical, associated with environmental sciences, agriculture, and enterprise development. Academic benefits would come with the intensive practice students would gain in sustaining visual attentiveness to illustrations and design, increasing vocabulary and reading abilities, and experiencing ways that learning in school translated to economic opportunities and local prestige. In addition, they would be immersed in meaningful roles of leadership and expertise that could enhance their future learning opportunities. New vocabulary would come in meaningful contexts and genres of speaking and writing not generally asked of the students in their classes during the year. Visitors to the garden would come from local governments, families that supported the school through philanthropy, and local business leaders eager to improve work opportunities for students within the school. When these visitors came to the garden, the students would be the spokespersons who explained the garden's purpose, design, planning, and eventual expansion. The students would have to answer technical questions and explain the relationship of the garden to the school's cafeteria and primary-level classes. Real audiences beyond school personnel would give meaning to the students' work as would the school's willingness for them to work with younger students in science projects related to the garden. We hoped that the roles taken on by the teenagers throughout the project would encourage them to become more accountable within their own school and community and more apt to be proactive in preparing to help the economic situations of their families and to advance their own prospects for career development in landscape design and agriculture.

Within modern Western history, literacy has been viewed as a set of skills and values essential to move along pathways to employment, creative industries, entrepreneurial enterprise, and property ownership. Our project's ideology reflected this set of expectations. Thus we saw the garden development, botanical illustration and photography, and school involvement for the adolescents as solid preparation for their possible futures. Our imagined futures for the adolescents saw them learning landscape design, advocating for community gardens available to the working poor, and using their skills and understanding of expenditures and income plus their reading and writing skills to build careers for themselves. At the school, the young people were to be fully involved in the financial planning and maintenance of the garden on a day-to-day basis.

Theories of Science Learning

This summer's pilot project was the beginning of what we expected to be a three-year full-scale development project to test in out-of-school informal environment, three theories of science learning previously assessed only in formal instructional or experimental settings. We devised for the D.R. adolescents a complex plan that integrated physical activity, visualization in creative learning, and science- and finance-centered communication with young learners in their school and outside experts. We saw the proposed design experiment as having strong potential for replication in urban communities of immigrant teens in the U.S. and in appropriate areas of Latin America. Three theories of learning were behind the design of the project.

The first theory argues that drawing and creating visuospatial representations advance learners' perception of details vital to seeing principles of science at work. The second focuses on the effectiveness of iconic gestures and enacted/

demonstrated roles in the learning of symbol structures that support logical–linguistic and numerical learning. The third argues the importance of identity alignment for motivating young people to imagine themselves as having a future that will be possible only with literacy. Applied within our project, this theory brought science and the basics of accounting together to enhance the career and life management possibilities of young people for whom the economic realities of their families generally precluded either science or basic financial planning. Our hypothesis in the pilot and looking ahead to the three-year design experiment was that these theories put into practice in informal environments would improve not only representational abilities but also conceptual depth and fluency in the genres and styles of talk expected of career-seeking young people. In particular, we were eager to enable the youth to see inside careers related to agricultural science and to understand the role of their own hard work and commitment as well as their ability to plan strategically and responsibly use both their time and money.

In the research we carried out during the pilot, we collected data on the teenagers' visual acuity for detecting the details within botanical illustrations, their own photographs of local flora and fauna, and written and illustrated information related to plants they selected for the school garden. We tracked their oral linguistic fluency and reading competency with materials related to the basics of community garden development and financial planning of an entrepreneurial project. Finally, we documented evidence of their conceptual grasp of core principles of science related to soil, composting, and plant development, as well as their ability to use their reading, writing, and accounting skills to record the garden's progress.

We collected baseline data on perception and representation of details in botanical illustrations in the early weeks of the project and at the end. Sobol and the team members (N, Z, P, and A) planned conversational prompts to elicit teens' envisioned scenarios of what the school garden could be. These audio recorded data enabled us to analyze teens' use of vocabulary and syntactic forms indicating sequence, cause and effect, and hypotheticals. We used a similar conversational prop at intervals to elicit any changes in the ways that teens expressed ideas about their possible futures. Throughout the project, we were focused on whether or not their participation in the project would lead them to think of themselves in new roles, such as agriculturalists, horticulturalists, visual artists, environmental advocates, entrepreneurs, educators, and family members aware of the importance of skills surrounding responsibility in keeping financial records and other written accounts vital to these roles.

Researchers in science education, as well as neuro and cognitive science, show the benefits of engaging young science learners in visual and spatial modes. Scholars have shown the development of creativity, memory, and linguistic fluency in scientific language that come with moves away from textbook and recitation learning (Ramadas, 2009). Visual thinking, mental modeling, and envisionment of act and consequence in future scenarios work in coordination with linguistic forms of reasoning (Roskos-Ewoldsen, Intons-Peterson, and Anderson, 1993; Suwa and Tversky, 1997). The high overlap in neural networks between visual perception of details and imagery suggests modal overlapping. Visuospatial thinking works as a mnemonic for verbal material and an aid to reasoning with mental models (Anderson, 1998; Schnotz, 2002).

In comparative experiments, imagery appears repeatedly to take precedence over language in developmental learning with regard to both maturational age and levels of conceptual difficulty of science information (Newcombe and Learmonth, 2005). Particular kinds of external representation (e.g. drawing vs. computer animation) facilitate conceptual change and skill gains better than others. Once visually perceived, details (whether of natural science scenes or diagrams, charts, illustrations, etc.) tend to transform as depictive and abstracted in working and long-term memory. However, the reverse is not necessarily the case: only in cases of direct training will purely verbal information call upon other representational modes or mental imagery and their correlated depictive and abstracting symbolic notations (Ramachandran and Hubbard, 2001). The sense of touch or haptic feedback gained through direct use of the hand appears to enhance the act of mentally visualizing (Reiner, 2008). Moreover, the hand as heuristic has long been attested (Wilson, 1998). Mental imagery helps discovery, creativity, and understanding in science learning (Ramadas, 2009; Stafford, 2007). The more directly the hand is involved—sketching, drawing, shaping—the more effectively mental imagery helps the learner to see ahead to envision future hypothetical scenarios, thinking through what could be and restructuring as needed. Sketching gives learners "perceptual interface" through which to discover underlying non-visual functional relations (Suwa and Tversky, 1997).

Young learners who engage with visuospatial and enacted representation undergo conceptual change that relies on coordination class (Di Sessa and Sherin, 1998). This view includes intuitive understanding and locally grounded experience as the productive foundation from which the learner integrates systematic and interlaced knowledge (Saxe, 1991; Sfard, 2008). The "funds of knowledge" of learners remain to be reorganized (instead of replaced) and amplified, modified, and theorized as learning advances (Gonzales, Moll, and Amanti, 2005; Vygotsky, 1978). To ensure identity alignment, young learners have to see something of themselves and their current interests in the modes, spaces, and models through which they learn science and mathematics (Greer, Mukhopadhyay, Powell, and Nelson-Barber, 2009; see also Gnijnik, this volume. For more on theories of learning behind the project, see Heath, 2011).

Failure Portended

In spite of all our theories, plans, and good intentions, the pilot project failed—at least in its long-term success, sustainability, and impact on learning. It failed in large part because of local cultural values surrounding class, color, and manual labor, as well as differences between assumptions about the benefits of young people taking on individualized responsibility in their education and career development in a nation uneven in its political operations as a democracy. We failed to include in our plans the barriers that projects based on values behind the "American Dream" face in a nation with the history—economic and political—of the D.R. We had also failed to anticipate the resistance within certain socio economic niches to the kind of garden and landscape design project we envisioned. A strong stigma is attached to the dirty, back-breaking work of planting, weeding, and harvesting monocrops such as sugar cane or pineapples. This work, in the view of the D.R. working poor, is what Haitians do. In the D.R., Haitians who have come across the border seeking work are hired to harvest sugar cane and other monocrops. Dominicans seek indoor work, such as jobs in hotels, factories, or small businesses in town. Many work for companies that serve the robust tourist industry of the D.R., as was the case in the city of our study, which is located near a major resort hub in the D.R.

The teenagers of the D.R. worry about the same issues of appearances as do their peers, who see the idealization of light skin on television and in films. In the D.R., distinctions among shades of skin color associate closely with agricultural work or other labor that requires working outside in the sun. In Dominican culture, light skin is often associated with beauty and, in many cases, wealth, so young Dominicans—particularly girls—try to stay out of the sun. Membership in the class structure of the D.R. tends to correlate with signs of "whiteness" and distance from forms of labor that expose the body to sun or hint association with cultivation of food, either in family gardens or in monocrop production. Many Dominicans at the upper levels of society not only reflect the greatest number of "white" features, but they also live and work at the farthest distance from manual labor and other conditions that could put these features at risk.

Those middle-class Dominicans outside membership among the wealthy include office workers, lower-level managers in multinational enterprises, including those in the tourist industry, and professional service workers, as well as small business owners and employees. These middle-class D.R. families provide their children primary and secondary schooling and some further education. Most own their own homes and cars. The Dominicans just below these middle-class families live in small homes away from the commercial center of the city or the residential areas of the wealthy.

The families of the young people for whom we had planned the project came from working-poor families, those Dominicans at the bottom of the social class ladder. Their homes are crowded, have sporadic electricity, and rarely have flushing toilets. Without ownership of their own cars, they rely on public transport to get to work and school, and their children have little access to medical or dental care or schooling beyond the elementary grades. The schools available to them are crowded and lack resources, including qualified teachers.

With keen awareness of the rules and norms of their social stratification, Dominicans of the working poor take great pains to distinguish themselves in every possible way from the very poorest residents of the D.R.—the Haitian immigrants. Though some Haitians have been living in the D.R. for several generations and have achieved some markers of middle-class status, their skin color, speech patterns and dialect, and "natural" association with their fellow Haitians at the bottom of the social rankings hold back their movement into leadership positions, citizenship, or social acceptance in the wider society. Most Haitians work in the fields and live in hovels that lie along the irrigation ditches of the fields. Schooling and medical care are almost entirely outside their reach. Charity work with Haitian children and their families, when it reaches the very poorest, often comes from short-term efforts of Christian groups from the United States. Handouts of clothing, food, medical care, and minimalist household goods mark the primary efforts of these groups, whose leaders may wish to do more to improve educational opportunities, but, when they try, barriers of the strict social hierarchy prove insurmountable.

Often with the support of multinational corporate philanthropy, local churches, and governmental funds, wealthy Dominicans have taken up the challenge of improving medical, dental, and educational opportunities for those Dominicans at the lower end of the social spectrum. Doing so requires not only a school but also medical facilities that intervene with children as early as possible—during the mother's pregnancy and throughout the elementary years. The school that was the site of our project was one of the few such facilities in the D.R., and the director of the school was carrying forward a vision that originated with wealthy families of her neighborhood in the 1970s. The school offered medical and dental care, daycare, pre-school, and, elementary classes for as many children as possible. Students were selected through lottery as well as family continuation. Workers in the school's cafeteria, as well as the janitorial and building maintenance staff, were parents of the children, and, to the extent possible, aides and technicians in the medical facilities and classrooms were also family members of the children of the school.

Social Capital

The school's social capital gave the project connections to governmental officials who agreed to supply seeds, visit the garden at its most flourishing state, and promote the school's garden project in the local paper. Benefactors and leaders of the school provided access to agricultural experts who agreed to help secure larger plots of land when the project was ready to expand. A local landscape designer and horticulturalist worked with the students and provided close professional contact for the young artists and gardeners. The social capital promise of the project was extended when a local businessman arranged scholarships for any graduate who decided to pursue higher education in agriculture management. Wealthy families who visited the project offered to provide apprenticeships with their regular landscapers and further work opportunities for those students who went on to set up their own landscape design enterprises.

At the conclusion of the summer's pilot, we had all felt success. Our early research indicated that during the pilot the students involved in the project had heightened their visual acuity, interest in agriculture, and sense of ownership over the garden. The teenagers had taken on several high-risk high-reward roles, and appeared confident and committed to the challenge of taking on significant responsibility. They showed great pride in the written manual they had created with illustrations from their own photographs and sketches. They were especially proud of the short documentary videos they had learned to make over the summer. On the last day of the project for this first summer, the students threw an "opening party" to show off the garden to their parents, teachers, and donors who supported the school. They spoke with confidence and enthusiasm about what they had learned and how much they felt they had grown in their sense of themselves and what they could do.

The research team left the D.R. with a sense of confidence that they had succeeded in working with the students to build their knowledge and skills in environmental science, selected art forms from photography to written expository prose, and public speaking. They seemed to have some sense of future career possibilities related to their development of the garden. They said "goodbye" to the team, knowing that Sobol would return within the next few months to visit them and the garden. Throughout the next year, Sobol returned several times and worked also with other college students from the U.S. who were volunteering in the school, some of whom agreed to help continue the work of the garden project during their stay in the D.R. In spite of this infusion of voluntary help by young people from the U.S., one of whom was originally from the D.R., the original vision, scheme, and energy of the project, as well as its ambitious set of goals, could not be revitalized. Each of Sobol's visits over that next year gave deepening evidence that the summer project shared some notable characteristics of a summer romance—brilliant while it lasts, but short-lived after the color, sights, and sounds of the summer fade away.

A year later, we had to agree the project had failed in creating any long-term impact on student learning or sense of commitment to either the school garden or the work and study necessary to take up any of the further educational opportunities that benefactors of the school had offered. The garden was still up and running, and its harvested vegetables were being used by the cafeteria, but the students were not among those who worked in the garden and watched over its cycles of abundance. The school's maintenance staff had had to absorb within their jobs responsibility for the garden and distribution of its produce. Almost as soon as Sobol, N, Z, P, and A left the D.R. after the summer, the original students lost interest and returned to their other extracurricular activities—baseball, basketball, and volleyball for some. Others went back to spending their half-days out of school helping with household work and childcare for their families.

What Happened?

The same factors outlined at the outset of this chapter regarding the dominance of class and social capital over literacy learning accounted to some degree for the failure of our project. However, in the end, for the young people the fundamental design of the project—based as it was in cultivating land for food, working outdoors (even in the cool hours of the day), and preparing to develop enterprises related to agriculture and landscaping—killed any prospects for continuation. Cultural values clashed resoundingly. Young men from the U.S. eager to advance sustainable community gardening, supplement D.R. teens' limited secondary education, and help students develop skills that would support a future with substantial job opportunities outside the tourist industry—all while putting solid learning theory to work—could not overcome cultural and political differences. Literacy learning, as a component of the project, could not in itself counter these factors.

Reading, writing, and financial accounting skills may enhance cultural capital, but these skills cannot in and of themselves provide sufficient social capital to overcome long-standing divisions based on attitudes toward ethnic or regional origin, skin color, and occupational niches of parents and prior generations. Structuring of symbol systems, such as alphabets, numeral systems, and the periodic table in chemistry, will not provide sufficient symbolic capital to overcome societal forces that block social mobility (Albright and Luke, 2008; Grenfell et al., forthcoming).

It is critical to note, however, that realities holding back the value of literacy acquisition for overcoming social and political forces do not spring only from groups divided by long-standing racial, class, and regional barriers operating at the community level in developing nations, such as the D.R. Restraining forces extend far beyond the local level to the heart of multinational corporations and their responsiveness to market forces. Market forces as well as shareholder demands for greater profit margins shape the current structural system surrounding wage earning for workers hired in extraction industries or agribusiness. Corporate forces are especially powerful in nations whose economy does not provide either sufficient education or manufacturing to support extensive employment beyond that of manual labor vital to extraction of natural resources and agribusinesses. Multinational farms and plantations dedicated to local crops such as rubber and sugar operate at the mercy of world markets where demand for these commodities is never stable. Synthetic alternatives may be developed at any time, either wiping out demand entirely or drastically reducing the price for the commodity on the world market. Such was the case with Latin America's rubber plantations by the middle of the 20th century when automobile tire production no longer relied on rubber trees. Introduction of aspartame, saccharin, and sucralose drastically reduced demand for sugar in modern economies.

Both international and local forces explain the low expectations for individual change of class status workers dependent on multinational corporations and tourism for employment hold. The most reliable aspects of their daily lives lie in the social relations and habituated patterns of their daily lives. To push beyond these is to undertake risk and to commit to a sense of personal agency as well as unfamiliar cognitive and linguistic demands. To learn to read, write, and calculate requires dedication and sustained cognitive and linguistic work that has little exchange value that the working poor can see. Instead, these individuals look to what they can see. In the case of the youth in the program we piloted, they saw their school and the services and benefits provided to the children and their families as requiring relatively little of the youth compared to the many demands the project was making on them. There they had to do manual labor, master and manage unfamiliar symbolic systems, and represent themselves to the outside world as having a future career beyond the school. The stretch was too much for the imagination to hang on to beyond the fun and novelty of a summer adventure with cool gringos from the North.

Lessons from the project are numerous. The teenagers involved did indeed show progress in discerning visual details, refining their drawings, gaining fluency in genres of explanation and demonstration, and becoming familiar with a host of new vocabulary items. Yet these gains had no possibility of continuation, for, when the school year started, the students became busy with their half-day classes that stressed rote recitation. Without the structure and programing facilitated by the project team, none of the students stepped up to continue playing the roles they had assumed during the summer—from project photographer to architectural designer of the garden gate. They did not see themselves as needing to remain committed beyond their diversionary involvement in the summer.

One additional factor contributed to the failure of our project. The teenagers we worked with are the chosen few of the working-poor, for they and their families have an unwritten promise of life-long connection to the school. The "natural" continuation of this privilege has come to be expected and accepted by them. One or more members of their families worked in the school, which meant that they had been accepted into the school through family ties. Many of the adolescents had younger and older siblings in the school. The encircling embrace of the school for families fortunate enough to be chosen extended from birth through secondary school and beyond. Jobs within the school were expected by those who had come through the school and whose family members had been long-term employees. In addition, the wealthy families who supported the school had a strong history of bringing "extra" material benefits to the school through volunteers from the U.S. The teenagers knew their future opportunities were in the good hands of the school's leaders and benefactors, and they were quite satisfied. They had to do little but be there and wait for the goodness of their benefactors to ensure their future. Developing an incentive to look for work beyond what was given could seem disrespectful as well as unnecessary. (Similar aspects of problems surrounding "giving to the poor" who see themselves as privileged enough has become a growing realization among social scientists and NGOs (non government organizations) around the world [cf. Dahl, 2009].)

Literacies, whether alphabetic or financial, gain and hold value only when they have an exchange value—what Bourdieu and others refer to as "symbolic capital" (Grenfell et al., forthcoming). Those who acquire literacies have talents and skills to give for which others exchange goods, services, jobs, and other forms of remuneration. For the D.R. teenagers, working for higher gains in reading, writing, and representing horticulture, gardening, landscape design, and small business accounting held no such exchange value; they did not have "symbolic capital," even though the outsiders helping them saw what they were learning as forms of "cultural capital." What they were given within their school would be enough for them to continue to have a place within the safety of their locale's encompassing environment.

Anthropologists since Margaret Mead have cautioned do-gooders, outsiders, and, most particularly, U.S. educators against assuming that what they value is wanted, needed, or will be accepted by others. From medical care to literacy to "improved" farming practices, Latin Americans have turned aside in order to maintain values close to them and their local situation. Every instance will vary slightly in circumstances and give different reasons for doing so. Yet certain generalizations are highly likely to hold regarding the risks of literacy in those situations where employment means crossing socio-economic class barriers supported by strong judgments about skin color, language, and types of occupation.

In the case of our project, we lacked sufficient cultural knowledge about local situations beyond the school to understand that raising food in gardens would not resonate with the students. They were unlikely to want to do such work as a career or to take part in the community gardens being promoted in some areas of the city. When the adolescents, the first secondary-level students the school had taken in, were asked about any dreams they had for the future, they named professions such as architecture that they thought would pay lucrative salaries. Social class and academic realities, however, would preclude such a career for any of these students. Not only did they lack sufficient math and science skills to pursue such a career, but they also lacked access to the "inner circle" of the wealthy elites who hired architects. Thinking of future careers meant for these students "moving up" in society, but they had little idea of what the paths to professional careers entailed and how social class arrangements surrounding higher education and enlistment of wealthy clients worked. As noted above, a patron of the school had arranged full scholarships for students to attend an agricultural management university upon graduation. According to the patron, pursuing a business in agriculture in the D.R. was a path that could yield substantial financial success for these students. He wanted therefore to ensure that those who dedicated themselves in the garden project had scholarships for going forward. But agriculture held no interest as a career for these students; that was an occupation meant for classes below them on the social ladder. Thus while agriculture might, in fact, have been a way they could advance themselves economically and break the cycle of poverty in which they grew up, they could not connect with images of themselves within agriculture. Instead, they held vague dreams of entering professions that would remain unattainable for them.

The problem in the case of our failed project was that the dream narrative of the professionals who founded and managed the school and those of us who planned and implemented the pilot program was not a narrative the young people could recognize. A central tenet guiding our project was to inspire and enable the students to become responsible change agents in their own lives and in their communities. Yet we were blindsided by the socio-political and cultural barriers that would prevent the teenagers from seeing any rationale for doing so. We could equip the students with certain skills, but the obstacles to real social mobility were far more powerful than ones we could overcome. The one feasible route the students could take to lift themselves out of their dependency on the school as their patron for life was that of learning the literacies that would enable them to pursue a degree in agricultural management. Yet the stigma associated with this choice of a future, compounded with the students' embedded notions of dream professions, overpowered choices that seemed eminently rational and sensible to us.

Reading and writing, handling financial planning and accounting, sustaining project work, and implementing ideas and dreams into action had little place within the guiding ideology of students and their families. Nor were competencies in multimodal literacies suitable for the working-poor of a nation firmly grounded in an economic and political history that prevailed to make unlikely the democratic political environment critical to redistribution of educational and socio-economic opportunities. Literacy theorists and practitioners have, for generations in all parts of the world, resisted the need to take into account practical means of managing ways to enable the poor to be able to count on literacy as having an exchange value sufficient to support social mobility. In countries where making any significant upward move in social class in the legitimate economy faces so many political obstacles, can literacy educators comprehend the risks young people must take if they learn to value, produce, and understand the symbolic structures that characterize multimodal literacies? Our research team learned through the harsh reality of the failure of our work that if we could not face this question, success by our definition would elude us as well as the young D.R. students around which the goals of our project were built.

References

Albright, J., and Luke, A. (2008). *Pierre Bourdieu and Literacy Education*. New York: Routledge.

Anderson, R. E. (1998). Imagery and spatial representation. In W. Bechtel and G. Graham (eds.), *A Companion to Cognitive Science*. Malden, MA: Blackwell, 201–11.
Dahl, B. (2009). The "failure of culture:" Christianity, kinship, and moral discourse about orphans during Botswana's AIDS crisis. *Africa Today* 56 (1), 23–43.
Di Sessa, A., and Sherin, B. (1998). What change in conceptual change? *International Journal of Science Education* 10 (10), 1155–91.
Gonzales, N., Moll, L. C., and Amanti, C. (eds.) (2005). *Funds of Knowledge: Theorizing Practices in Households, Communities, and Classrooms*. Mahwah, NJ: Erlbaum.
Greer, B., Mukhopadhyay, S., Powell, A. B., and Nelson-Barber, S. (eds.) (2009). *Culturally Responsive Mathematics Education*. New York: Routledge.
Grenfell, M., Pahl, K., Rowsell, J., and Street, B. (forthcoming). *Language, Ethnography and Education: Bridging New Literacy Studies and Bourdieu*. New York: Routledge.
Heath, S. B. (2011). "If you don't see it, you don't get it!" Seeing our way into learning science in informal environments. In W. Tate and C. C. Yeakey (eds.), *Research on Schools, Communities, and Neighborhoods: Toward Civic Responsibility*. Washington, D.C.: American Educational Research Association.
Newcombe, N., and Learmonth, A. (2005). Development of spatial competence. In P. Shah and A. Miyake (eds.), *Handbook of Visuospatial Reasoning*. New York: Cambridge University Press, 213–56.
Ramachandran, V. S., and Hubbard, E. M. (2001). Synaesthesia: a window into perception, thought, and language. *Journal of Consciousness Studies* 8 (12), 3–24.
Ramadas, J. (2009). Visual and spatial modes in science learning. *International Journal of Science Education* 31 (3), 301–18.
Reiner, M. (2008). The nature and development of visualization: a review of what is known. In J. Gilbert, M. Reiner, and M. Nakhleh (eds.), *Visualization: Theory and Practice in Science Education*. Surrey, U.K.: Springer, 25–9.
Roskos-Ewoldsen, B., Intons-Peterson, J., and Anderson, R. (1993). *Imagery, Creativity, and Discovery: A Cognitive Perspective*. Amsterdam: Elsevier.
Saxe, G. (1991). *Culture and cognitive development: studies in mathematical understanding*. Mahwah, NJ: Erlbaum.
Schnotz, W. (2002). Towards an integrated view of learning from text and visual displays. *Educational Psychology Review* 14 (1), 101–19.
Sfard, A. (2008). *Thinking as Communication: Human Development, the Growth of Discourses, and Mathematizing*. New York: Cambridge University Press.
Shavelson, R., Phillips, D., Towne, L., and Feuer, M. (2003). On the science of education design studies. *Educational Researcher* 32 (1), 25–8.
Stafford, B. (2007). *Echo Objects: The Cognitive Work of Images*. Chicago: University of Chicago Press.
Suwa, M., and Tverksy, B. (1997). What architects and students perceive in their sketches: a protocol analysis. *Design Studies* 18, 385–403.
Vygotsky, L. S. (1978). *Mind in Society*. Cambridge, MA: Harvard University Press.
Wilson, F. R. (1998). *The Hand*. New York: Vintage Press.

10

THE BRAZILIAN LANDLESS MOVEMENT AND A MATHEMATICS EDUCATION RESEARCH PROGRAM[1]

Gelsa Knijnik

Introduction

This chapter analyzes a Mathematics Education Research Program developed with the Brazilian Landless Movement in the south of this Latin American country in the last 20 years. Even though the research program is focused on the study of different fields of mathematics (one of them being school mathematics) and the analysis of school mathematical discourse, this program can be considered as part of Latin American Literacy Studies. In fact, we are taking as the theoretical framework an ethnomathematics perspective, which is based on a social-cultural approach that understands the learning of mathematics as acquiring and using mathematical languages (in plural, as we will discuss below), more precisely, networks of mathematical language games, in Wittgenstein's terms. This implies considering that each network is composed of socio-cultural practices, which lead us to situate the Mathematics Education Research Program in the field of Literacy Studies.

The Research Program is part of a worldwide theoretical approach started in the 1970s, which highlighted the cultural dimensions of mathematics education. During that time the mainstream field of mathematics education was impacted by a new perspective. Different socio-political processes (struggles for democracy in countries like Brazil and struggles for independence in colonial countries like Mozambique) were seen to establish the *conditions of possibility* (Foucault, 2002) for the emergence of this cultural approach in mathematics education. The new world socio-political configuration shaped the need to diminish the high rates of illiteracy and the large number of children of those countries who failed in mathematics at school. In Brazil, the work of Freire (2006) in what is called, in Latin America, *Educação Popular* or *Educación Popular*, the studies developed by Nunes, Schliemman, and Carraher (1993), and the ethnomathematics field established by D'Ambrosio (2006) are examples of social-cultural perspectives in mathematics education which emerged during that period.[2]

In the last decades, studies based on the theoretical framework of ethnomathematics have been developed at research centers in Latin America, North America, Asia, and Africa (Barton, 2004; Duarte, 2009; Gerdes, 2010; Giongo, 2008; Hirsch-Dubin, 2009; Knijnik, 2006, 2007; Moreira, 2006; Powell and Frankenstein, 1997; and Wanderer, 2007. The notion of culture is at the kernel of these studies whose main goal is to identify and analyze non-mainstream Western cultures, highlighting their mathematical ideas. Using Wittgenstein's later work, it can be said that their main goal is to identify and analyze non-hegemonic forms of life and the mathematical language games to which they are associated.

Since it began, the Research Program coordinated by Knijnik (2009) has developed studies in the field of ethnomathematics focusing on rural areas of the southernmost Brazilian state, Rio Grande do Sul. This was associated with the beginning of the work with peasants of the Landless Movement (in Portuguese, *Movimento Sem Terra*) which had emerged a few years before, precisely in that part of the country. The importance of this social movement in Latin America, especially as regards their schooling processes, allowed us to develop ethnomathematics studies which shaped the Research Program discussed here.

The chapter is organized in four sections. The first is a brief description of this peasant social movement and the Educational Project associated with it. The second section focuses on the theoretical framework of the Research Program. In the third, methodological aspects of the Research Program are described and empirical data are analyzed. The last section of the chapter focuses on the pedagogical repercussions of the Research Program.

Brazilian Landless Movement and its Educational Project

In recent years Brazil has achieved an outstanding international position thanks to its political contribution to the worldwide scene and its economic and social growth. The World Human Index shows that among the Latin American countries it currently occupies eighth place. Even so, 10 percent of the population controls 75 percent of the national wealth and the 2010 National Census shows that 16 million people live in miserable conditions, earning less than U.S. $40 a month. The country also has the second-highest concentration of land ownership in the world; 84.4 percent of the population live in urban areas; and 14.6 million who are older than 10 years are still illiterate: 21.3 percent of those living in rural settings and 6.8 percent of the urban population.

But in the mid 1980s, when the Landless Movement (LM) started, the country's political, social, and economic situation was even worse. As a result of the military dictatorship (1964–1985), when profound political and economic changes occurred in the country, such as the rapid industrialization and mechanization of agriculture, millions of rural Brazilians moved to the urban areas, a social shift that began to exert a profound influence (see also Castañheira, this volume). One such influence was the LM. Since its beginning, its main struggle was for agrarian reform. Following the fall of the dictatorship in the mid 1980s, the new Brazilian Constitution established that land must have a social function and required that the Brazilian government "expropriate rural properties that are not performing their social function for the purpose of land reform." In the last 25 years, since it began, in order to address these issues, the LM has achieved a growing political position in the country. Gathering hundreds of thousands of peasant families, the LM trajectory shows the importance of ensuring education for children, youths, and adults in their communities.

Currently, the LM runs approximately 2,000 public schools in encampments throughout Brazil; more than 200,000 landless children and youth are attending school; since the inception of a National Literacy Project in 1993 more than 50,000 peasants have become literate; the LM runs Teacher Training Courses attended by over 3,900 educators at secondary and higher levels and 250 pre-school centers with children from 0 to 6 years old; and has established partnerships with more than 13 public universities for bachelor's and master's degree courses in Education, History, and Administration, among others. The Movement also runs "itinerant schools," which follow occupations of land by formerly landless people—the parents of the pupils at these schools and other families of rural communities who were looking to better their economical and social conditions affected by the continued process of Brazilian land concentration. The schools were established more than ten years ago, and today there are more than 45 operating in 7 Brazilian states. Approximately 350 landless teachers and more than 4,000 students work at these "itinerant schools," which follow the encampments settled by land occupation.

The data shown above render explicit the dimensions of the Landless Movement Educational Project of which schooling is a part. According to one of the LM's official documents, the schooling project sees the "need for two articulated struggles: to extend the right to education and schooling in the rural area; and to construct a school that is in the rural area, but that also belongs to the rural area—a school that is politically and pedagogically connected to the history, culture, social and human causes of the subjects of the rural area (...)" (Kolling, Cerioli, and Caldarte, 2002, 19).

The LM has dedicated itself to conceiving the schooling of its children, youths, and adults by paying attention to these two struggles. This means that landless educators consider peasant culture a key issue also for those teaching and learning processes involving mathematics. But they explicitly mention that the value ascribed to it cannot deny the relevance of acquiring mathematical tools connected to mainstream school mathematics. They consider that these tools can improve the use of new technologies for managing production in rural areas and can allow the learners to go further in their schooling trajectory. As will be shown in the next section, these ideas are strongly connected to the ethnomathematics perspective that underlies the Research Program we have been developing with the landless peasants of Southern Brazil.

Theoretical Framework of the Research Program

The theoretical framework of the Research Program developed with the landless peasants[3] is aligned with post-metaphysical theoretical positions. To them, the problems "that afflict humanity need to be considered in the world, in history, in the harshness of the concreteness and the practice, without appeal to transcendental concepts, to essences" (Ribeiro, 2010). Specifically, the Research Program is supported by Michel Foucault's thinking and Wittgenstein's later ideas, which are expressed in his book *Philosophical Investigations*. According to such theorizations, we conceptualized the ethnomathematics perspective as a toolbox, which allows the analysis of the Eurocentric discourses of academic and school mathematics and their effects of truth; issues of difference in mathematics education, taking account of the centrality of culture and the power relations that institute them; and the language games that constitute different mathematics and their family resemblances.

At this point it is necessary to argue about the theoretical consistency of putting together Foucault's and Wittgenstein's perspectives. I consider that this consistency can be achieved through the language concepts adopted by both philosophers and their non-metaphysical positions towards knowledge. Moreover, Wittgenstein's notion of language games can be linked with the Foucauldian notion of "truth games" (Castro, 2009, 423).

Differently from what Wittgenstein proposed in the *Tractatus logicophilosophicus* (1994), in his work of maturity, *Philosophical Investigations*, when he sought to answer the question: "What is language?" he says that "we must not ask what language is, but rather how it functions" (Condé, 1998, 86). When operating this theoretical shift, Wittgenstein indicates that it is no longer possible to talk simply in a language, but rather in languages, i.e. "a huge variety of uses, a plurality of functions or roles that we could see as language games" (Condé, 1998, 86). Therefore, the meaning of a word emerges as we use it in different situations.

Hence the same expression, in different contexts, will mean something different. In Wittgenstein's words: "the meaning of a word is its use in the language" (Wittgenstein, 2004, 20). As we can observe, Wittgenstein abandons any essentialist concept of language (a principle that we now also apply to literacy). Indeed, if the meaning of a word is determined by the way we use it, use can be understood as something that determines a practice and not "as the expression of a metaphysical category" (Condé, 2004, 48).

Thus, Wittgenstein's theoretical production in his maturity and some of his interpreters such as Condé (1998, 2004) and Glock (1996) allow the inference that language games and the rules that constitute them are strongly affected by the way we use them. This means that language games should be understood as immersed in a form of life, strongly amalgamated with non-linguistic activities. Glock argues that "a form of life is a culture or social formation, the totality of communal activities into which language-games are embedded" (Glock, 1996, 125). Indeed, since meaning is given by use, the meaning can change at every use we make of the words. "What we do is to bring words back from their metaphysical to their everyday use" (Wittgenstein, 2004, 48), to the friction of "rough soil."

These ideas lead us to the notion of form of life as "the intertwining of culture, world-view and language" (Glock, 1996, 124). In this intertwining, the meanings we give words are mediated by rules that are conceived in our social practices. A set of such rules constitutes a grammar that, as indicated by Condé (2004), is very important in order to analyze modern rationality, because it "guides" the interactions between the different language games. Underlying the emphasis on learning to operate with the rules of grammar, Condé says that Wittgenstein means "grammar and language games as a rationality that is forged from the social practices in a form of life and that is no longer based on ultimate principles" (Condé, 2004, 29). When one abandons the idea of a single, natural, reason-producing structure, it is possible to understand rationality as an "invention," a "construct." It is this "construction" that will allow language to articulate itself inside a form of life and establish how rationality will indicate to us what we should accept or not. It is important to highlight that the philosopher's notion of "family resemblance" also points to the possibility of interconnections between different language games belonging to the same form of life or to different ones.

Thus, when pointing out that two language games have family resemblances, this does not refer to an identity between the games. It is emphasized that both have similar aspects and that they are distributed by chance, without a supposed uniform repetition. These ideas point to a non-essentialist theoretical perspective which is shared by Foucault's positions. In short, Wittgenstein's philosophy of maturity, denying the existence of a universal language, enables us to question the notion of a universal mathematical language. This allows us to argue, from the philosophical viewpoint, about the existence of different mathematical language games: those learned and taught in the school form of life and those practiced outside school forms of life. Following Wittgenstein's ideas, we can consider that both kinds of mathematical language games have family resemblances. Foucault's theorizations enable us to analyze these different language games in their "ties with the production of power-knowledge relations and with constituting truth regimes" (Wanderer and Knijnik, 2007, 3). The theoretical framework of the Research Program briefly discussed above, notably the plurality of different mathematical games and their relation to issues of power, has been productive in our mathematics education work with the landless peasants in two different dimensions. These will be discussed in the next section.

Mathematics Education Work with the Landless Peasants

In the last 20 years the landless settings of the Research Program have included schools at encampments and settlements, adult education projects and teacher education courses at secondary and bachelor levels, mainly in the southernmost state of Brazil.

Fieldwork has been done, taking as reference postmodern anthropology, specifically its ethnographical perspective. Therefore, methodological procedures—such as participant observation, audio recording, field diary, and interview—have been used. Even considering that these methodological procedures are not ethnography in the strictest sense—rather, as Green and Bloome (1997) argue, they involve "adopting an ethnographic perspective"—we were very much aware of the need to take into account questions and challenges which have been asked contemporarily by anthropology, an area strongly marked by its links with the colonial era, with the description of the "Other"; we are now more systematically "reflexive" about such positions (Foley, 2002). From these perspectives, we must consider the power relations that are exercised during fieldwork and also how we describe in our papers what happens there.

An example of a work in which these methodological procedures were used was developed in a mathematics education curriculum subject at a landless teacher education course.[4] There I discussed the specificities and family resemblances of three different mathematical language games with the students. The first was a language game produced by a form of life associated with landless peasants (characterized by the Brazilian rural culture of small land properties but, in this case, with big influence of LM multidimensional struggle for land reform. The settlements and encampments are democratically structured and in most of the settlements the production is organized cooperatively). The second was produced by a form of life of the male urban sawmill workers (characterized by the culture of small villages of the southern state of Brazil, an urban way of life which still maintains some rural traditional values and habits). And a third language game was produced by a form of life taught in Western Eurocentric schools (which were introduced in the country by the Portuguese colonization in the beginning of the 16th century).

The two first language games analyzed with the students referred to the "cubagem" of wood (in Portuguese *Cubagem da Madeira*—cubic meters of wood) in order to calculate "how many cubic meters[5] there are in a truckload." This procedure is a common practice in landless culture. The peasants perform it when it is necessary to build houses or animal shelters in camps and settlements, and to purchase or sell planks, i.e. "in our negotiations with the sawmill men," as one landless peasant said. Throughout my work with the students I realized the importance they give to such language games, produced by their form of life. Particularly in teacher education courses and at settlement schools I have found great interest in discussing that practice, constituted by a specific grammar, a specific set of rules (in the sense taken by Wittgenstein, as referred to earlier).

In the school mathematics classes they were expecting to go further in their learning about the cubagem of wood language game. But it was expected that I also assumed another role. In consonance with the LM pedagogy (Chassot and Knijnik, 2008), they aimed to acquire the knowledge of school mathematics: the one they called "book mathematics." Avoiding a naïve perspective, they were aware of the social importance of such a language game and they needed to learn its specific grammar as part of their struggle for agrarian reform (see also Fuenlabrada and Delprato, this volume).

Even considering the theoretical difficulties involved in translating language games from one form of life to another, here I will express the landless language game of cubagem of wood using words and the syntax of the school mathematics language game. I am aware that in doing so, some (or maybe most of the) specificities that constitute the landless' form of life which produced the cubagem of wood language game are suppressed. So, it can be said that the language game here shown became a "hostage" to the school mathematics language game when it is said that it basically involves two steps: the first one was to identify, by modeling, a tree trunk with a cylinder whose circumference coincided with that of the middle part of the trunk. The second was also identified by modeling the cylinder in a quadrangular prism, whose measure on the side was one fourth of the perimeter of the cylinder base. In other words, in the first step a tree trunk (with its irregularities of width) was "geometrically simplified." It was transformed into a cylinder with the following characteristics: its base was defined by measuring the circumference measure of the middle part of the trunk (which, in general, is an approximation of the average of the different widths of the trunk); and its length was defined by the length of the trunk. In the second step, the cylinder built in the first step was itself simplified. Now, the cylinder is transformed into a quadrangular prism: its square base is obtained by considering as the side of the square a fourth part of the measure of the cylinder circumference. Thus, this language game finds, as trunk volume, the volume of the quadrangular prism whose side of the base was obtained by determining the fourth part of a circumference. This, in turn, corresponds to the cylinder base, obtained by modeling from the initially given tree trunk.

But another language game emerged in that pedagogical process. At some point in the discussions, a two-student dialogue produced a shift in the debate about the first language game learned.

Student A: The measurement process that I know is almost the same [as the one learned before], except that we measure at the narrow end of the wood.

Student B: The point is that the right thing would be to do it in the middle. But the purchasers do not want to buy a piece that will fall away, if they want it for square wood [wooden plank] or things like that. They will want a piece that goes from here to there [which goes from one end to the other of the log]. Those chips that are produced will only be for burning.

According to these students, there were urban sawmill men who did not use the "middle of the log" as reference, considering only its narrower end, since they were interested in obtaining whole planks. For this purpose a different rule of calculation was introduced, conforming to a specific grammar, which leads to a new language game, different from the first one. The students called the language game practiced by sawmill workers who live in the urban areas close to their communities "the sawmill method". We found that we were dealing with a language game which is produced by a specific form of life, different from that of the landless peasant.

The pedagogical process was not limited to those two language games. The language game associated with the one taught by school mathematics (named by the students as book mathematics) with its specific rules that shape its grammar was also analyzed. Moreover, the family resemblances of these three language games—the landless, the sawmill, and the school language games—were emphasized.

The work involved studying the modeling process of the landless language game and learning mathematical tools such as relations between a cubic meter and its multiples. In different situations the results of calculating the "amount of wood" obtained by their language game were empirically compared by the participants to the volume of the cylinder produced by using part of their language game, which would correspond, in their view, to a better approach to the total quantity of wood of the trunk, reckoning not only the part useful to obtain "whole planks," as in the school language game. The students also found that the results of their language game minimize those obtained using the cylinder volume. Nevertheless, the group showed particular interest in learning "the formulas of book mathematics" connected to the discussion we were holding. They argued that these abstract formulas could be useful to them in the future (to continue their schooling process and/or to participate in teacher public exams that will allow them to obtain a permanent public teacher job in the educational system). In learning how to calculate volumes of the cylinder and rectangular prisms (which also requires knowing how to determine the length of the circumference, the area of a circle), the group was dealing with the specific grammar which constitutes the language game of Western Eurocentric school mathematics.

Bringing those three language games into the mathematics class enabled the group to go further in the appropriation of rules that shape the rationality produced by each of those forms of life. They learned more about the landless peasant cubagem of wood practiced in their communities. When the family resemblances of those language games were analyzed, the students were able to identify the "remnants" of wood that were produced by their language game, which were even greater when the initial measure of the log circumference was determined at the "narrow end," as considered by the sawmill people language game. So, in this case, the wood not used for making planks could be useful for other purposes and therefore, in given situations, it should also be included in the accountancy of their calculations.

Summing up, it can be said that learning about different mathematical language games and their family resemblances allowed the peasant students to broaden not only their mathematical world but also their ways of seeing the complex social relations involved in different forms of life that produce such different language games. But it was not only the landless peasants who had learning experiences. As a researcher, I learned important lessons. They taught me about the need to mobilize all my efforts in order to never forget to problematize my own discourse, since it is necessarily marked by my "privileged" voice as an intellectual working with non-hegemonic social groups. In my attempts I have been favored by this social movement, which is very much aware of the risk of exposing themselves to academic research and of being narrated by "the others," of being represented by us.

The LM chooses to take this risk not only because it knows that I am careful about my role in the work I have been developing with them. It is also because they see education as one of the central issues of their struggle and consider the importance—at least at this point of their trajectory as a social movement—of having scholars contributing to the construction of their educational schooling processes, to their teacher education courses, which, for them, are key elements for the social justice project they are attempting to build. But here a point must be clarified: to be accepted as contributors to the LM schooling processes and/or to its teacher education courses, the academics must show, before starting their work, the theoretical convergence they have with LM educational perspective. Moreover, throughout their contributions the movement develops a permanent evaluation, which is decisive to how they define the continuity of such contributions.

Pedagogical Repercussions of the Research Program

In carrying out our research program we are facing issues connected to politics of knowledge, to the dispute around the definition of which knowledges are included and which are excluded in the schooling processes (see Hernández, this volume). This dispute is marked by power–knowledge relations, which ultimately legitimate and are the legitimizers of some discourses, which interdict others, precisely those that are about the knowledges, the rationalities, the values, the beliefs of cultural groups we place in the position of "the others." LM educational perspective aims to subvert the politics of dominant knowledges; it does not assume a naïve position connected to an exacerbated cultural relativism. This is not the case. But the "price to be paid" for acquiring the hegemonic school mathematics subjects cannot be the silence of peasants' knowledges in the school curriculum.

One should then ask how a single rationality among other rationalities, for example, the rules by which individuals and cultures deal with space, time, and quantification processes—all that which Western civilization associates with the notion of mathematics—became a "truth," the only "truth" that could be accepted as mathematics in the school curriculum. What is at stake here is to problematize the sovereignty of Modern rationality, which looks down on all other rationalities associated with "other" forms of life; the existence of a single mathematics—"the official one" —with its Eurocentric bias and its rules marked by abstraction and formalism. School mathematics—the traditional set of knowledges taught at school—inherits at least part of the formal and abstract grammar that constitutes academic mathematics, through pedagogical recontextualized processes, in Bernstein's words (1997).

In summary, it can be said that the set of language games of academic mathematics, as well as those of school mathematics, offers a "dream of order, regularity, repeatability and control … and with it the idea of a 'pure', disembodied reason" (Rotman, 1993, 194).

The ideas presented in this chapter about a Mathematics Education Research Program developed with a Brazilian peasant social movement may inspire other Latin American countries as well as hegemonic ones. The old myth of "homogeneous schools" is being destroyed by the new world configuration. In recent decades, migratory movements have increased worldwide, and many immigrant children are now at school in central countries such as those belonging to the European Union and the United States. The field of education and particularly mathematics education has gained awareness of the need to pay attention to this socio-cultural phenomenon that has concrete repercussions in schooling and broader social contexts (Abreu, 2001; Civil, 2007). Now the foreigners are "here." What kind of rationality operates in their social practices? How do these practices differ from the school mathematics rationality? More than ever these questions have become central to education.

Notes

1 Mathematics education studies also assume a social–cultural framework, and their authors characterize them as Quantitative Literacy (Steen, 2001) or Numeracy (Ferreira, 2007; Coben, 2008). These studies aim to highlight their mathematical dimension. Nevertheless, this nomenclature does not conflict with the broad meaning ascribed in this book to Literacy Studies.
2 Out of the Latin American context the work of Gerdes (1991) in Mozambique, Bishop (1988) in Pacific Islands and Zaslavsky's studies (1973) in African countries must also be mentioned.
3 Here it should be mentioned that empirical data which have been produced by the set of studies that constitute the Research Program are not restricted to the landless peasants. Other cultural groups have been involved, like rural German-descended communities and communities of Technical Agricultural schools.
4 What follows is a summarized version of the paper: Mathematics education and the Brazilian Landless Movement: three different aspects in the context of the struggle for social justice (Knijnik, 2007).
5 The terms "cúbicos" and "cúbicos de Madeira" are used in the Brazilian rural areas to mean cubic meters of wood. The term "metros de Madeira" —in English, "meters of wood" —is also used.

References

Abreu, G. (2001). O papel mediador da cultura na aprendizagem da matemática: a perspectiva de Vygotsky [The role of culture as mediator in learning mathematics: the perspective of Vygotsky]. *Educação, Sociedade e Cultura [Education Society and Culture]* 13, 105–17.

Barton, B. (2004). Mathematical discourse in different languages: implications for mathematics teachers. In B. Clarke, D. M. Clarke, G. Emanuelsson, B. Johansson, D. V. Lambdin, F. K. Lester, A. Wallby, and K. Wallby (eds.), *International Perspectives on Learning and Teaching Mathematics Teaching Mathematics*. Göteborg, Sweden: National Center for Mathematics Education, 365–78.

Bernstein, B. (1997). *La estructura del discurso pedagógico: clases, códigos y control, Vol. IV* [*The Structuring of Pedagogic Discourse, Vol. IV: Class, Codes and Control*], Pablo Manzano, trans. Madrid: Morata.

Bishop, A. (1988). *Mathematical Enculturation: A Cultural Perspective on Mathematics Education.* Dordrecht: Kluwer Academic Publishers.

Castro, E. (2009). *Vocabulário Foucault* [*Foucault's Vocabulary*]. Buenos Aires: Biblos.

Chassot, A., and Knijnik, G. (2008). L'Education dans le Mouvement Sans Terre: la construction d'une pédagogie par un mouvement social. [Education in the Landless Movement: A pedagogy construction by a social movement], in J. Houssaye (ed.), *Pédagogues de demain?* [*Pedagogies of Tomorrow?*]. Cedex: Université de Rouen, 307–27.

Civil, M. (2007). Building on community knowledge: an avenue to equity in mathematics education. In N. Nassir and P. Cobb (eds.), *Improving Access to Mathematics: Diversity and Equity in the Classroom*. New York, NY: Teachers College Press, 105–17.

Coben, D., and Hodgen, J. (2008). Assessing numeracy for nursing. In M. Joubert (ed.), *Proceedings of the British Society for Research into Learning Mathematics* 28 (3), November 2008, 18–23.

Condé, M. (1998). *Wittgenstein: linguagem e mundo* [*Wittgenstein: Language and World*]. São Paulo: Annablume.

Condé, M. (2004). *As Teias da Razão: Wittgenstein e a crise da racionalidade moderna* [*The Intricacies of Reason: Wittgenstein and the Crisis of Modern Rationality*]. Belo Horizonte: Argvmentvm Editora.

D'Ambrosio, U. (2006). *Ethnomathematics: Link Between Traditions and Modernity*. Rotterdam, Netherlands: Sense Publishers.

Duarte, C. G. (2009). *A "realidade" nas tramas discursivas de Educação Matemática Escolar* [*The "Reality" in the Discursive Plots of School Mathematical Education*]. Unpublished doctoral dissertation, Universidade do Vale do Rio dos Sinos, São Leopoldo RS, Brasil.

Ferreira, M., Magalhães, M., and da Penha, M. (2007). Parâmetros para avaliação de habilidades matemáticas dos alunos em iniciativas de alfabetização de jovens e adultos. [Parameters for the evaluation of Mathematics students in adults and youth education initiatives.] *Revista Iberoamericana sobre Calidad, Eficacia y Cambio en Educación* 5 (2e), 232–40.

Foley, D. (2002). Critical ethnography: the reflexive turn. *Qualitative Studies in Education* 15 (5), 469–90.

Foucault, M. (2002). *Em defesa da sociedade: curso no Collège de France (1975–1976)* [*In Defense of Society: A Course at the Collège de France (1975–1976)*]. São Paulo: Martins Fontes.

Freire, P. (2006). *Pedagogia do Oprimido* [*Pedagogy of the Oppressed*], 46ª ed. São Paulo: Paz e Terra.

Gerdes, P. (1991). *Etnomatemática: cultura, matemática, educação* [*Ethnomathematics: Culture, Mathematics, Education*]. Maputo: Instituto Superior Pedagógico.

Gerdes, P. (2010). *Da Etnomatemática: a arte-design e matrizes cíclicas* [*From Ethnomathematics: to art-design and cyclical matrixes*]. Belo Horizonte: Autência.

Giongo, I. M. (2008). *Disciplinamento e resistência dos corpos e dos saberes: um estudo sobre a educação matemática da Escola Técnica Agrícola Guaporé* [*Discipline and resistance of bodies and knowledge: a study about mathematical education at the Technical Agricultural School Guaporé*]. Unpublished doctoral dissertation, Universidade do Vale do Rio dos Sinos, São Leopoldo, RS, Brasil.

Glock, H. (1996). *A Wittgenstein Dictionary*. Oxford: Blackwell Publishers.

Green, J., and Bloome, D. (1997). Ethnography and ethnographers of and in education: a situated perspective. In J. Flood, S. Heath, and D. Lapp (eds.), *A Handbook of Research on Teaching Literacy through the Communicative and Visual Arts*. New York: Simon and Schuster Macmillan, 181–202.

Hirsch-Dubin, F. P. (2009 October). Mayan Elders, Mayan Mathematics, and the Weaving of Resistance in Maguey Bag Production. *The Journal of Mathematics and Culture* 4 (1), 63–83.

Knijnik, G. (2006). *Educação matemática, culturas e conhecimento na luta pela terra* [*Mathematical Education, Cultures and Knowledge in the Land Struggle*]. Santa Cruz do Sul: EDUNISC.

Knijnik, G. (2007). Mathematics education and the Brazilian Landless Movement: three different mathematics in the context of the struggle for social justice. *Philosophy of Mathematics Education Journal* 21 (1), 1–18.

Knijnik, G. (2009). Ethnomathematics research in a postmoder time. In K. Selati, R. Vithal, C. Malcolm and R. Dhunpath (eds), *Researching Possibilities in Mathematics, Science and Technology Education*, 1a. ed., Vol. 1. New York: Nova Science Publishers, Inc, 37–53.

Kolling, E., Cerioli, P., and Caldart, R. (eds.) (2002). *Educação do Campo: Identidade e Políticas Públicas* [*Education in the Country: Identity and Public Policies*]. (Coleção Por Uma Educação do Campo, n. 4.) Brasilia, DF: Articulação Nacional por uma Educação do Campo.

Moreira, D. (2006). Texto matemático e interacções [Mathematical text and interactions]. In G. Knijnik, F. Wanderer, and C. J. Oliveira (eds.), *Etnomatemática, currículo e formação de professores* [*Ethnomathematics, Curriculum and Teacher Training*]. Santa Cruz do Sul: EDUNISC, 432–46.

Nunes, T., Schliemann, A., and Carraher, D. W. (1993). *Street Mathematics and School Mathematics*. New York: Cambridge University Press.

Powell, A., and Frankenstein, M. (eds.) (1997). *Ethnomathematics: Challenging Eurocentrism in Mathematics Education*. Albany, New York: State University of New York.

Ribeiro, T. M. (2010). *Do "você não pode" ao "você não quer": a emergência da prevenção às drogas na Educação* [*From "You Cannot" to "You Don't Want To": The Urgent Need for Drug Prevention in Education*]. Unpublished master's thesis, Universidade do Vale do Rio dos Sinos.

Rotman, B. (1993). *Ad Infinitum: The Ghost in Turing's Machine*. Stanford: Stanford University Press.

Steen, L. A. (ed.) (2001). *Mathematics and Democracy: The Case for Quantitative Literacy*. U.S.A.: The National Council of Education and the Disciplines.

Wanderer, F. (2007). *Escola e matemática escolar: mecanismos de regulação sobre sujeitos escolares de uma localidade rural de colonização alemã do Rio Grande do Sul*. [*School and School Mathematics: Regulation Mechanisms on the School Subjects of a Rural Location of German Colonization in Rio Grande do Sul*]. Unpublished doctoral dissertation, Universidade do Vale do Rio dos Sinos, São Leopoldo.

Wittgenstein, L. (1994). *Tractatus Logico-Philosophicus*, 2a ed. São Paulo: Edusp.

Wittgenstein, L. (2004). *Philosophical Investigations*. Oxford: Blackwell.

Zaslavski, C. (1973). *Africa Counts: Number and Pattern in African Culture*. Boston: Pindle, Weber & Schmidt.

11

READING, WRITING, AND EXPERIENCE: LITERACY PRACTICES OF YOUNG RURAL STUDENTS

Gloria Hernández

During my contact with the reading and writing practices of rural youth in a community in central Mexico,[1] I have found that the experiences that young people have constructed with writing throughout their lives acquire a distinctive place in the way they position themselves in relation to literacy at school. This positioning surpasses the mere referential function announced through many of the textbook activities that are used in the classrooms: whilst teachers do in fact often ask students to point out what they know about a given subject, based upon their own experience, the students' own social practices involving literacy provide not only for the construction of knowledge, of the kind the teacher is asking for, but for broader understandings related to the shape and contexts of use in which this knowledge is constructed.

In this chapter, my intention is to approach the ways in which these young rural students construct their literacy experiences, as well as the potential for these experiences to be used within the school context. Youth studies tend to focus more on urban environments; therefore, since the rural milieu has not been sufficiently explored, a central aspect for the development of this analysis is the rural condition of these young students. The chapter is a subproduct of an ongoing study and it is based on the experiences of my own teaching practice developed for research purposes. I chose the cases of two students for analyzing the construction of their literacy experience, and, for the purposes of this chapter, I will call them Mario and Pedro.

These boys attend a small school with eleven other students. The thirteen middle school students, who are 12 and 13 years old, have had at least six years of literacy contact through their schooling. This has introduced them to a wide spectrum of possibilities for using literacy and to the likely pleasures of reading and writing, although the young people's attitudes towards these two activities are not always positive or homogeneous. Some value literacy from the perspective of having the obligation of using it in the ways that school requires them to, as the way of having the option to continue studying and remaining in school, while others do not. Half of the students declared that they read and write at home so as "not to get bored." Their situation of extreme poverty prevents them from having Internet access, going to the movies, having other recreational activities, or buying books, making writing a possible recreational activity.

Generally, the extrapolation of students' experiences and knowledge to the classroom is limited, because, given the school codes, the opportunities for bringing out-of-school knowledge to the classroom fails to take into account the ways in which students construct their experience with literacy. After teaching Mario, Pedro, and the rest of the students in this class and looking into their worlds and ways of life, I can advance the hypothesis that their literacy experience is integrated by a legacy of previous knowledge, as well as by the processes, contexts, conceptions, and interactions through which they acquired it. Additionally, precisely this experience goes beyond what the formal school classroom is asking for, attaining rather the shape of a "potential movement," defined by Dewey as the impulse that moves an individual towards curiosity, strengthens his (or her) initiative, and creates desires and purposes (Dewey, 2004, 81).

The need to recover students' experience is a widespread practice, acknowledged in textbooks as well as through teachers' classroom interventions. However, this recognition only refers to what learners know about a topic, but not to the ways in which they have arrived at knowing what they know, or the diverse contexts in which they use that knowledge. As a result, despite the appearance of approaching a more complex knowledge, there is in fact only a sort of artificial recovery of previous ideas and conceptualizations, in which students simply repeat in chorus for their teachers whatever it is that they had learned before, rather than the "potential movement" of which Dewey speaks.

What is common in the classrooms I have studied as well as in others reported in the literature (Kalman and Carvajal, 2007) is the triggering question "What do you know about this?" But not "How do you know it?" or "What do you think about it?" or "Where did you learn it?" or "With whom did you learn it?" With this, only two things are strengthened: on the one hand, repetition without reflection about the constructed knowledge; and on the other, the notion of experience only as a product, and not as a process with multiple interactions and wrapped in larger complexities.

Another important element in the classroom recovery of students' experience is that it is selective and literacy-related knowledge is discriminatory. This was already explained by Gee (2005, 39), who, without making any reference to students' experiences, refers to the competences privileged by school, and by doing this he places the subject of power use within the recovery of the overall classroom experience.

When I write about "experience" I do not only refer to past events evoked in the present, but to a complex relationship that places these adolescents before a travelled road as readers and writers. This is an intricate relationship that also places them in their present and in what they perceive as their future within those roles. This is illustrated when they state, for example, "I have never liked and I will never like to read or to write." It is precisely this kind of relationship that is the central theme of this chapter: the various ways that the literacy experiences and practices in young people's everyday contexts influence their reading and writing practices at school.

This chapter has two main parts: first, a close-up look at the case studies, and then a broader discussion of the middle-school students' use and understanding of reading and writing in and out of school. Using Mario and Pedro as telling cases, the discussion will include commentary and analysis of the students in general.

The Community as a Reference for Structuring Literacy Experience of Young People: The Cases of Mario and Pedro

The context of this project is a *Telesecundaria*[2] located at a rural community in central Mexico. From the official perspective, a rural community is defined as an inhabited area of maximum 2,500 inhabitants, who are mostly involved in agricultural activities. However, for the purposes of this chapter, I prefer to use the definition proposed by Rosas: rural communities are the ones defined by their occupation, their social organization, their community dynamics, and the values that guide their lives (Rosas, 2005, 294).

The community is small: it has an elementary school and the *Telesecundaria*. The latter is the highest level of schooling available in this locale, and for those students who cannot travel to a neighboring town to continue their studies, it is the highest institution of formal schooling that young people can aspire to attend. Thus, some teachers call the *Telesecundaria, La máxima casa de estudios* (the highest learning institution) in analogy to the way the National Autonomous University of Mexico (UNAM) is often identified throughout the country in official discourse.

For men in the community, agriculture is the main occupation, although some of them are also employed as construction workers. Sometimes women work in the fields during the sowing season, but for the most part domestic chores and child rearing are their main occupations. The activities of young people differ: boys generally help in planting and harvesting corn and in taking care of the animals while girls usually help with the housework; yet taking care of small children is a task carried out by young boys and girls alike. Reading and writing practices are, as we will see, entwined within specific demands of these occupations and within family situations.

For the parents who can send their children to this school, the fact that they finish middle school has great relevance, but they do not really place importance on their children's compliance with school rules. As a teacher, I have become aware of parents' strong negotiations with us so that we do not ask students to do a lot of homework. They argue that children get tired and have other things to do, like helping in the fields or taking care of their young brothers and sisters or nephews and nieces, and that they do not always have time to be with books in their hands doing nothing or wasting time. The image of reading as doing nothing means that they are not doing anything for the benefit of urgent domestic chores or productive jobs. The value of going to school is seen in the future and not within immediate activities, and it is centered more on the individual than on the collective. Consequently, when grownups consider that their children should be working, they see reading as a waste of time.

But by being close to the students, I observed that because they do not have movie theaters or other recreational activities, reading offers a possibility for having fun. In the context of reading a book of short stories that was recommended and loaned to him by a classmate, one student said that reading was a way of not getting bored, but when, he said, "I don't understand the words, I throw away the book and I do nothing. I get desperate."

Despite the fact that this rural town has a community library with specialized books for young people,[3] their experience tells them that the use of the library is associated with homework, and therefore they do not visit it for recreational purposes. They only go there when their teachers tell them to go and never to look for recreational reading. They also have a classroom library with 30 books, but they do not use it because there is no scheduled time for that purpose: it needs to be set up by the teacher, and during all the years I have been in the school, this has never happened.[4] The school library also has an important collection of specialized titles for students their age; nonetheless, this library is never used either. For all these reasons, contact with books is limited to the official textbooks that students receive free of charge as part of the *Telesecundaria* program, and to the ones that are used within the classroom context. For the most part, students do not have access to other books; during their out-of-school literacy practices they produce their own texts, like verses, songs, short stories, messages, graffiti, and letters. More than the readers of texts written by others, they are the writers and readers of their own written productions. They write poems and short stories, they write down the lyrics of the songs that they listen to on the radio, and they create their own songs.

All the first year middle school students have had at least six years of literacy contact through their schooling. This has introduced them to a wide spectrum of possibilities for using literacy and to the likely pleasures of reading and writing, although the young people's attitudes towards these two activities are not always positive. They value literacy from the perspective of having the obligation of using it in the ways that school requires them to and as the way of having the option to continue studying and remain in school.

What follows is a description and analysis of the life contexts of Mario and Pedro and of their uses of writing, their experiences at home and at school, and the relationship between their daily practices in literacy. The first of the two is Mario, a young man dedicated to his family with an important role in its wellbeing; the second is Pedro, a dedicated student and budding writer. Both come from the same community, live in similar situations of poverty, and yet their experiences are quite different. These two cases are examples of the general trend observed at the *Telesecundaria*.

Mario's Literacy Experiences

Mario is a 14-year-old boy who every day takes care of three small siblings under the age of 5 for several hours in the morning. This requires that he uses several literacy practices related to either reading instructions on medication labels, or writing the notes that his mother—who cannot write—dictates to him regarding medication schedules and the ways in which medications or household remedies need to be administered to his sick younger brothers. In Mario's construction of this experience, the elements that come into play are a conception of himself as a responsible person, his love for his smaller brothers, the acceptance of the role that he assumes within the family context, and, as he puts it, the need to overcome "the fear of making a mistake and that something happens to my little brothers."

Mario is also the one in charge of reading the texts on fertilizer bags, and this is another important way that he contributes to his family for planting and growing corn. Printed on the sacks are instructions that need to be followed and carefully applied in the family's management of their fields in order to have a good crop and to produce enough corn to be able to both sell some at the market and keep the surplus for their own consumption.

The structuring of these experiences with literacy has its basis in a family commitment, in the value of childcare, in the value of planting, in being responsible, and in belonging to a family collective within the framework of an extended family. Mario's family is composed not only of parents, brothers, and sisters but also of aunts and uncles, nieces, nephews, cousins, and in-laws.

The ways young people like Mario live their youth are anchored in these roles. Since from their families' point of view they are not children anymore, they can and must develop those everyday chores that will prepare them to become someone in life, and not only the ones necessary for doing well at school. Within this context, the construction process of Mario's literacy experience goes back at least six years, because he entered elementary school late and started reading at eight. Then, at home, he approached reading through the above-described documents, and since his mother does not know how to read or write she delegated these activities to him, recognizing the value of use of her son's reading and writing abilities. Another circumstance contributing to Mario's taking on so many responsibilities is that his father recently emigrated to the United States looking for work.

Contrary to other young people who write as a way of constructing their identities (Hernández, 2009), Mario does not write much because he feels that he does not need to do so. He only writes within the school context or following the instructions his mother dictates to him, because he already takes on the responsibility of reading as a way of helping his family. He does not go to the community library to do his homework, where he could have more contact with books, as some of his friends do. While he believes that reading instructions for medications satisfies a family need, he considers that reading books is boring and unnecessary. However, initially, he did not really consider that reading those medication labels was reading at all, because he fully adhered to the homogenizing school conception that disregards other literacies as legitimate practices with their own meanings and forms of organization, different from the ones promoted by schooling.

As has been mentioned in other studies, the concept of reading that Mario had is consistent with the one that is transmitted via school culture and other means: that reading is only related to reading books, and that these books should be legitimate; in other words, validated by the power of school culture, or "good" books, assuming that there is a shared vision of what this means (Sarland, 2003, 38). What is being ignored is that this criterion for what counts as a book is acquired within the book's reading context of use and it is not in the book per se. This reading concept is also set up within Mario's experiences of literacy practices, given that the potential that reading might have is virtually unknown to him. What is important to him is the success he has in the contexts of use related to the different roles that he plays and to the meaning reading has for him.

When asked about a particular topic from school, Mario has a hard time responding even when the subject is directly related to his experience as the caretaker of his younger siblings. While at home this activity is of great relevance to him and to the family collective; at school it embarrasses him, and for that reason the value and the uses of his reading materials go unmentioned. The contexts for reading at school and constructing reading experience and the context for reading at home take separate paths due to the intrinsic meaning they have in each specific space. However, the difference is not only established by the contextual specificity in which the reading experience is constructed, but basically by the impossibility of dialogue between both contexts. This impossibility is established by the school's legitimating power, a power through which the social character of literacy is defined from a singular vision; or, in Street's words, from the pedagogization of literacy (Street and Street, 2004).

Mario has a second-hand mobile phone that he bought with the few dollars that his father sent him. He seldom uses it for talking or for sending messages because he has no money to buy credit, but, conscious of the status symbol of this artifact among young people, he does not miss the chance to display it when he is with them.

Writing on the mobile phone follows the codes that are currently used by young people: they only write the main consonants of a word: q for *qué* (why/what), or use mathematical signs =, ÷, ×, + to write short texts. These signs are used to represent the words equal to, among, by, or more. For Mario, this is the language of the youth, *los chavos*:[5] his friends understand it, and by using these symbols instead of words he can save money on the amount of words needed to send a message. This usage of literacy is linked to his condition of being young within his community and is consistent with the fact that young people have very little money and cannot afford to put a lot of credit on their mobile phones. Being young demands specific writing strategies for strengthening this youthful condition. Within this writing experience, Mario plays another role: here it is the role of a young person who, consistent with his interests and possibilities, uses the language and the symbolic gadgets of his generation—the cellular phone—instead of pencil and paper.

Pedro's Literacy Experiences: Context and Interaction

Pedro's mother is a housewife who finished elementary school and his father just returned from the United States. During his stay in that country he used to call his son and tell him that, if he had a problem or needed to talk to a man, he should approach his school teacher. Pedro has great recognition for the school and for his teacher; the family atmosphere always encouraged a sense of appreciation for school and convinced him of the importance of complying with school assignments and not indulging in some of the pastimes that some of his friends did. Upon his return to Mexico, Pedro's father showed concern about him—of his two children, Pedro is the only son he has. He attends parents' meetings at school and pays attention to Pedro's school work.

The context of this 13-year-old boy is one of very strong family support. His family treats him as a child still, and, although Pedro participates in housework and in planting activities, his main responsibility is to study to get good grades at school and to be a good boy. This means that he cannot participate with other young people in forbidden activities like hanging out, stealing, or drinking alcohol.

Since Pedro lives in extreme poverty, he produces his own short stories to fight boredom, and, when his mother sends him to the store, he keeps the papers that the products he purchases are wrapped in because usually these are pieces of newspaper that he reads to extract "some words that I need to write my stories." In other words, he looks for more vocabulary. He uses the free textbooks that are at his disposal at the *Telesecundaria* for the same purpose. With these resources, and making use of his great imagination, Pedro makes up his own stories. These are very long and are usually divided into chapters that end with the phrase "to be continued," as can be seen in the following text.

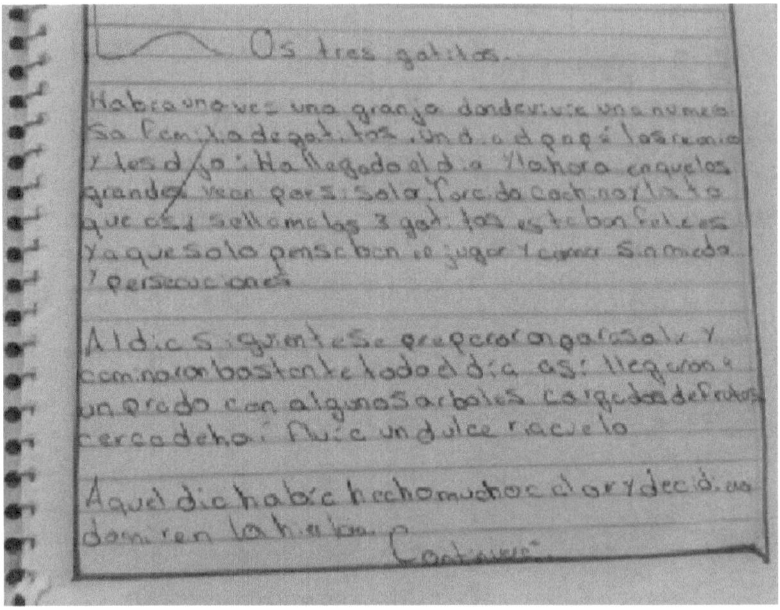

FIG. 11.1
The three little kittens.
Once upon a time there was a farm where a nume
rous family of kittens lived. One day the father got them together
and told them: The day and the time has come for the big ones
to look after themselves. Crooked one, dirty one and clever one
those were the names of the three kittens they were happy
because they only thought about playing and eating without fear
or persecutions

The next day they prepared themselves to go out and
They walked a lot all day long until they reached
a meadow with some trees full of fruits
close from there a sweet streams a flowing
It had been very hot that they so day decided
to sleep on the grass

To be continued...

It is striking that Pedro does not know what will be the end of the story, and that he creates a sense of mystery and expectation. Because he studies in the evening, this mystery is his motivation to wait eagerly for the mornings to arrive so he can continue writing, making time between doing housework and helping his family in planting corn. These writing practices give him the possibility to learn about the conventions of writing, such as punctuation and spelling; but also this experience directs him to the very act of writing, of imagining, of reading for writing, of creating mysteries for himself, instills a sense of fun, and is a way to spend time, and to learn more and more words.

All this provides Pedro with a potential experience that gives him several elements: he is—he says—his best author, because he can have the pleasure of manipulating the stories and the characters; of changing the way the story ends by proposing three different alternatives and choosing one; to laugh at what the characters do, to think all night about what the end will be, and to make "[the characters] do very difficult things that I cannot do." He cannot buy story books and he has no access to books at school or at the community library; therefore, he produces his own. *A day in the life of Pedro* is reflected in the following text:

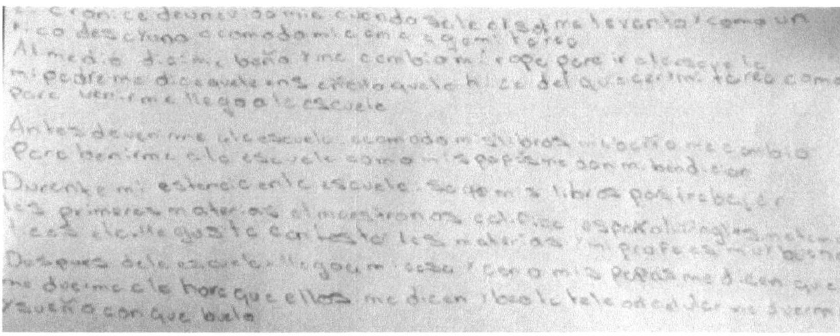

FIG. 11.2

Chronicle of a life of mine. When the sun comes out I get up and eat
a delicious breakfast I make my bed I do my homework
At noon I take a bath I change my clothes to go to school my father asks me to show him
what I did for him as work and my homework I eat
to come I arrive at school
Before I come to school I put my books in order I take a bath I change my clothes
to come to school I eat my parents give me my blessing
During my stay at school I take out my books to work
the first subjects the teacher grades us [are] Spanish English mathema
tics etc. I like replying to the subjects and my "profe" (short for teacher) is very kind
After school I arrive at my house and have dinner my parents tell me to
Sleep at the time they tell me I watch TV. or the cellular I sleep
And I dream that I fly

As it can be observed, the production of a text does not impose on Pedro the spelling rules applied at school. His text presents several problems (such as accentuation and word spacing) despite the fact that he is one of the most advanced students in his class. However, getting high grades does not necessarily mean that he has an academically strong appropriation of orthographic and syntactic conventions of school writing. In contrast to Mario, for Pedro, doing homework, arranging books, and his father's monitoring of his schoolwork, structure a central element in his experience with literacy that in this case is directly associated with his school experience. At home, his parents reinforce this hegemonic idea of writing within the framework of the meaning and commitment that they believe that school has for their son's education; therefore, what allows him to produce texts is not the appropriation and excellent use of spelling conventions, but the framework of high recognition to the school that his parents transmit to him day after day. This is what structures the construction of his experience with writing and what consolidates the hegemonic notion that he has of the vision of writing that comes from the school context.

In Pedro's experience the teacher is not only a mediator that brings him closer to the texts but an authority figure, given the fact that his father emigrated to the north, in search of work. The interaction he has with his teacher comes from an authentic affection for him: for Pedro, the teacher is his guide, and someone who reads and appreciates his stories, because he is one of the few people who has access to them. And although he does not correct directly his spelling and grammar, this access allows the teacher to play a warm mediating role between Pedro and his own texts, and between Pedro and other texts. Thus, the teacher's mediation in Pedro's interaction with written texts goes beyond a pedagogical approach to books; this mediation is constructed within a web of interactions and care in which the job of the mediator reaches its highest meaning (when he actually brings Pedro close to the texts of other authors for learning purposes), which is the case with the textbooks. But, more importantly, he values Pedro's stories, and through that the boy consolidates himself as an author.

In this particular case, the emotional interaction adds to the usual mediation between the subject and the text; and at the same time this emotional mediation is mediated by specific conditions within a context in which the image of the teacher represents not only a pedagogical authority, but also an educational authority in charge of Pedro's education in the broad sense of the term. Thus, the mediation of the teacher is inserted into a cultural pattern; he is a man, he is caring, and he is someone to whom Pedro can turn whenever he needs a sympathetic ear or an authority figure to guide and advise him. He is also the one who teaches him school contents, and with all that, in Pedro's opinion, he is the one who teaches him all he needs to know in order to be a good person in life. On occasions, this interaction with Pedro is mediated by writing, by the reading of textbooks, and by support for the production of his own texts. On others, the interaction is only sustained by the communication and the emotional bond that they have between them. This significant interaction also structures Pedro's experience with the written word to the extent that he links the accompanied use of writing with being a better person. The appropriation experience of the written word is, as noted by Kalman, "part of a power relationship and the way in which people appropriate it depends on the prevailing social and cultural practices and not only on pedagogical and cognitive factors" (2008, s/n). As noted by Heath and Sobol (this volume), cultural practices do not always lead to appropriation of literacy or even an appreciation of it. This helps explain the difference between Mario's instrumental views of reading and writing and Pedro's interest in literary creation, even though they come from the "same" community.

In Pedro's literacy experience, the role of the mediator suggests that, although the intervention of the teacher helps him approach the texts of others, it also gives him author's support through accompaniment and through the teacher's appreciation for the texts he produces; an element of significant experience that may be part of that "cultural capital that one must possess in order to produce reading situations" (Colomer, 2002, 14).

With Pedro's literacy experience, I see that this is not just a legacy that can be put into a functional use in the present, but that it also refers to the action of experiencing writing, which in itself is significant, and, above all, to the transformative value of the experience within Pedro's own concept of writing, and of himself as author. Thus, although experience is a continuum, this does not mean that it is linear. Before anything else, to the extent that it allows our positioning before something new that makes us learn other things in a different way, experience is otherness (Larrosa, 2009).

The group of students with whom I work display a great deal of experience with the written word, not only at school, but via the community social practices in which they may or may not employ all of the formal conventions of written language. The production of texts in this classroom is a practice that was held during several occasions throughout the school year, but, in that school context, it was linked to evaluation and it had strict timing and specific rules. Both the school and the community are part of Pedro's literacy experience, but they differ in their context of use and the literacy experience in itself is hardly changed by the school from the process that Pedro develops beyond school. Pedro's experience has modified his practices:

> Before [he refers to two years ago], I used to write with spelling mistakes. I wrote a lot. Now I am better at spelling and you can understand better [what I write], I write stories, homework, school assignments, and copies of books. I do all that because I have to, because it is homework, but the stories, I write them because I get bored.

The progress that Pedro refers to, and that he considers as an improvement regarding good spelling and which allows him to write more, does not necessarily imply that he rigorously adheres to the rules, as we can observe in his story. However, it does imply a better use of the conventions in terms of the progress he has had. He values his own achievements in the construction of his experience with average writing, even if in the eyes of the school logic he still has strong spelling mistakes.

In this comment, Pedro's out-of-school—and out-of-boredom—literacy experiences, as well as in-school practices, are articulated. While creativity prevails in the first, in the second what prevails is their compulsory character; copying instead of creating. Unlike Mario, for whom the community appears as the favored element that benefits from his reading of instructions in medication labels and fertilizer packages, for Pedro, what seems to structure his writing for fun is his own world. His family is the collective that strengthens and monitors school uses at home; that is, the entity that reinforces the association of literacy with learning and pedagogy (Street and Street, 2004, 204).

Experience as a Potential Tool for Writing: Evidence from a Middle School Classroom

The value that Mario's and Pedro's classmates give to reading and writing is related to their memories about each school level.[6] While in preschool, literacy activities were linked to play, and were more appreciated by the *Telesecundaria* students. At elementary school, reading and writing turned into a series of packaged and repetitive activities—answering written questions that were posed by the teachers and not the students, taking exams, reading out loud in front of the group (an obligatory activity that they seldom enjoyed), and doing "lectura de comprensión," reading done for the purpose of identifying the main ideas in a text—all according to the teacher's idea of what a main idea constitutes. And, at the middle school level, for some students, reading and writing got "more and more difficult every time."

Even though these difficulties are linked to learning contents that consider gradual increases in complexity, they are also associated with the lack of configuration of an experiential continuity (Dewey, 2004, 77) in favor of the students' present reading and writing practices. Building on this notion, experiential continuity assumes that not every experience is valuable in itself or for the subject's future development within his or her interactions with literacy. Valuable experiences refer to a degree of distinction in regard to what the subject had been doing, or to the way in which he or she had been interacting. This is why, Dewey adds, "Every experience is a moving force. Its value can only be judged on the ground of what it moves toward and into" (Dewey, 2004, 81).

The recognition of the diversity in the reading and writing practices of these young men in and out of school suggests that experience plays a crucial role in their learning processes as readers and writers and that this experience is made up of a complex construction process of literacy practices that involve dynamic transformations not only of reading and writing, but, as I have already pointed out, of the concepts of reading, writing, and self, intervention of others, the uses that they make of reading and writing, and the objective treatment of these perspectives in the production of written texts through school codes and power relations.

The limited recognition by the school of these boys' experiential knowledge is related not only to the lack of acknowledgment of their lives but also to the lack of appreciation of their experience and the structuring processes that result from who they are and what they do during their contacts with literacy. Although in Mario's case his reading experience is structured from the responsibility roles that he plays within a particular collective and through specific actions like growing corn—or from the respectable self image that he has of himself as a responsible young man who is capable of giving support to his family group—at school he defines himself and he is defined by his teacher as the worst reader, as someone irresponsible for not doing homework, and as a problem student without any possibility of improvement.

In sum, the writing and reading experiences out of school of the young students in this rural community have little or nothing to do with school—where expressions of boredom are punished—and, in any case, where writing is used not to overcome boredom, but to punish it even further, i.e. by filling page after page with: I must be silent. Consequently, the uses of writing—for having fun, for expressing and communicating feelings, for being creative, for participating in family life, and for executing punishment—differ, and their meaning is contrasting for the students. However, together they constitute the reading experience of these young boys and they both structure and give content to their understanding.

Recognition of previous knowledge should be understood not only in terms of specific school content knowledge, but should also include an appreciation of the literacy broader experiences of these young rural boys. This would allow for their development in terms of the construction of their awareness and ownership of valuable practices of reading and writing as well as their mobilization of school contents. I think that this is another function of the teacher as mediator if he or she does not want to depart from the conception of the students as an empty vessel that Paulo Freire wrote about (1976). It is in this sense that experience can be transformed into a potential tool, and not into a teaching gimmick that ignores the contexts in which it was created. To visualize this literacy experience at school allows learners not only to bring prior knowledge into the experience, but by bringing experience into their knowledge, the boys' understanding of themselves as producing subjects with many opportunities for interacting with the written word would be enhanced.

Notes

NOTE: The author of this chapter has used pseudonyms throughout to protect the identities of students and instructors.

1 This paper is part of a project I started three years ago named "Training in science and literacy" and its general purpose is: to have an impact on the learning improvement of the students who are enrolled in "*Telesecundaria*"; from the development of field studies in the area of training in science and literacy, to the development of alternative resources to improve teaching practices in the above-mentioned fields of study. It is important to point out that throughout the research design phase I was authorized to work as a teacher in that program and that I systematically taught two subjects one or two days per week: Spanish and biology. And although these activities allowed me to develop certain actions and more closeness to the group, they were also developed within the framework of the timing and decisions established by school dynamics and the strategies of the teacher who was responsible for the first grade group. This teacher imparts all the subjects in accordance to the *Telesecundaria* model in Mexico.

2 In Mexico, the *Telesecundaria* was created in the decade of the 1960s to eliminate the educational gap in rural, indigenous, and semi-urban areas that had a television signal. It is a schooling modality that started transmissions in 1968 and has had an important growth until the present. Today *Telesecundarias* represent 51.7 percent of the total of middle schools in the country and they serve a total of one fifth of the six million students registered at this level of education. (INEE, 2007, 95).

3 The community library has a collection of 2,030 books with titles related mainly to school topics. There are children literature books as well as titles with topics of interest to young people. There are also books with other topics such as cooking and herbal medications.
4 As I mentioned before, the teaching activities that I carry out are developed within the framework of the school management regulations and the group dynamics developed by the responsible teacher. In other words, I work as a teacher within the limits of the establishment and I only have incidence in certain activities.
5 In Mexico the word *chavo* defines a young person or an attitude associated to being young.
6 In Mexico, basic compulsory education is composed by preschool (three years), elementary school (six years), and middle school (three years).

References

Colomer, T. (2002). El papel de la mediación en la formación de lectores [The role of mediation in the training of readers]. In *Consejo Nacional para la Cultura y las Artes, Lecturas sobre Lecturas* [*National Council for Culture and Arts, Readings about Reading*] , Vol. 3. México: CONACULTA, 9–29.

Dewey, J. (2004). *Experiencia y educación* [*Experience and Education*]. España: Biblioteca Nueva.

Freire, P. (1976). *Pedagogía del oprimido* [*The Pedagogy of the Oppressed*]. México: Siglo XXI Editores.

Gee, J. (2005). *La ideología en los discursos* [*Ideology in Discourse*]. Madrid: Morata.

Hernández, G. (2009). Identidades juveniles y cultura escrita [Youth identities and literacy]. In J. Kalman and B. Street (eds.), *Lectura, escritura y matemáticas como prácticas sociales: Diálogos con América Latina* [*Reading, Writing and Mathematics as Social Practices: Dialogues with Latin America*]. Mexico City: Siglo XXI Editores, 186–201.

Kalman, J. (2008). Alfabetización, diversidad y poder: las lecturas y las escrituras [Literacy, diversity and power: readings and writings]. In *Centro de Cooperación Regional para la Educación de Adultos en América Latina y el Caribe, Alfabetizaciones* [*Center for Regional Cooperation for Adult Education in Latin America and Carribean, Literacies*]. Pátzcuaro, Michoacán, México: CREFAL, s/n.

Kalman, J., and Carvajal, E. (2007). Hacia una contextualización de la enseñanza y el aprendizaje en las aulas de la Telesecundaria [Towards a contextualization of teaching and learning within Telesecundaria classrooms]. *Revista Latinoamericana de Estudios Educativos* [*Latin American Journal of Educational Research*] XXXVII (3–4), 69–106.

Larrosa, J. (2009). Palabras para una educación otra [Words for another education]. In L. J. C. Skliar (ed.), *Experiencia y alteridad en educación* [*Experience and Otherness in Education*]. Argentina: FLACSO Argentina y Homo Sapiens editores, 189–203.

Rosas, L. (2005). Una mirada a la educación rural [A glance into rural education]. In B. Teresinha (ed.), *Anuario Educativo Mexicano: Visión retrospectiva* [*Mexican Education Annual: A retrospective view*]. México: Universidad Pedagógica Nacional y Miguel Ángel Porrúa, 293–302.

Sarland, C. (2003). *La lectura en los jóvenes: cultura y respuesta* [*Reading Among Youth: Culture and Response*]. México: Fondo de Cultura Económica.

Street, J., and Street, B. (2004). La escolarización de la literacidad [The schooling of literacy]. In V. Zavala, M. Niño Murcia, and P. Ames, (eds.), *Escritura y sociedad: nuevas perspectivas teóricas y etnográficas* [*Writing and Society: New Theoretic and Ethnographic Perspectives*]. Lima, Perú: Pontificia Universidad Católica del Perú, Universidad del Pacífico Centro de Investigación, 181–210.

12

TECHNOLOGY AND LITERACY: TOWARDS A SITUATED COMPREHENSION OF A MEXICAN TEACHER'S ACTIONS

Irán Guerrero

Introduction

Nico is a seventh-grade geography teacher. Since the 2006 curriculum reform in Mexico for the middle-school level, he faces growing demands from local authorities to use technology when he teaches. For him, the computer is not "the solution" to existing problems, but an opportunity to look for information and project it, or "to do something" with the images. As a way of responding to this new requirement, and to figure out what that "something" may mean, Nico has designed situations that promote the use of the computer by his students. He adds, "there are other teachers worse than me" (EN2180907), referring to the fact that this technology is not readily available to all teachers and students (Kalman, 2004).

In this chapter I analyze the use that this teacher makes of technology (the computer), and how such use is situated in a complex context shaped by social forces such as the technology-related policies in schools and classroom conditions. For this, I first present the theoretical and methodological background; then, I refer to some of the ideological constructions that are adjacent to the policies that promote the use of technology in Mexican classrooms. In the next section, I describe a complex classroom context in which such teaching practice is developed; and in the fourth and last section I present an account of how the teacher in this case study uses technology. Throughout this unfolding analysis, my intention is to characterize the ways in which the educational reforms and policies that promote the use of technology-related innovations in the classroom are inserted into well-known arrangements in which, although the possibility of change exists, this change may not be as radical as expected:

> Considered as bricolage, innovation in local social practice (including discourse as the practice of talk) can be seen as tactical reuse of pre-existing

> elements. As a process of hybridization it innovates by connecting the old with the new—that is precisely how it accomplishes the new. What is "new" in such innovations is only a tiny percentage of the whole. What makes for novelty is the new combination of forms to perform certain functions, not the newness of the individual formal elements themselves. The new juxtaposition of elements of form within new arrays of functions makes for a change from the way in each element or feature is seen separately. This is a kind of reframing.
>
> *(Erickson 2004, 167)*

The changes proposed by the 2006 middle school reform are neither automatic, nor radical, nor provoked by the presence of technology. I will argue that the use of technology for teaching purposes in this context implies the use of literacy practices and demands the recognition, as well as the reintegration, of certain contextual elements. As my example will demonstrate, the use of technology involves social practices that can potentially promote the construction of knowledge and meanings and therefore cannot be isolated from the contexts in which these are produced. Technology use in Mexican schools, then, requires a deep analysis of the way this use can contribute to or restrict the students' and teachers' learning opportunities.

Methodological and Theoretical Points of Reference that Outline the Study of Technology Use

This chapter is part of a broader research project.[1] Fieldwork took place in four secondary schools during the school year of 2006–2007. I observed the teachers' work in the Media Room (a room with computers and a person in charge of managing the hardware, software and didactic material), as well as in each of the teachers' classrooms. I chose Nico's class as a case study of how one teacher uses technology in his teaching and designs a situation for his students to use the computer and the Internet to complete an academic assignment using texts and images. To carry out the analysis, I first reproduce the conditions of the classroom in which this teacher's actions are situated and modified; and in order to achieve this, I use dimensions such as time, spaces, actors, interaction, and feelings to examine what I observed (Spradley, 1980).

In order to analyze the interaction in the classrooms I took an *ethnographical approach* and made use of elements from sociolinguistics and the New Literacy Studies (Heath and Street, 2008). My central interest was to study some specific aspects of everyday life in the classroom and the social practices that take place there (Green and Bloome, 1997). This approach helped me bring out other important contextual elements; understanding these as what accompanies, gives shape to and situates the production of meaning by the actors (Floriani, 1994). From sociolinguistics I incorporated approaches to the analysis of speech, that involve inferring the teacher's thinking and understanding and the socio-pedagogical links among the different actors (Cazden, 1988; Coates, 1996; Edwards and Mercer, 1987; Gumperz, 1999; Kalman, 2005; Mercer, 1997).

Literacy is more than knowing how to read and write. From the New Literacy Studies perspective (Street, 2008), its meaning refers to how people "participate in the social world through written culture, without invalidating the different ways and formats that are used for this purpose" (Kalman, 2003, 17). The texts' nature and genre, as well as the needs to produce meaning, have been modified with technology (Ferreiro, 2010; Kalman, 2009) (an email is different from a letter or a telegram), and they need more complex perspectives for their analysis:

> Literacy is often narrowly considered as simple acts of communication through the use of written and read words, and sometimes includes speaking and listening. Thus literacy "skills" and instruction focus on the description of these skills and school activities to deliver their intentional development. But literacy is far more, and to examine its nature and consequences within technologies of communication requires a broader, cultural perspective.
>
> *(Myers, Hammet, and McKillop, 2000, 102).*

The use of communication and information technology can be understood as an extension of literacy because it encourages the use of different modes of representation—such as sound, image, or video—to build meanings. In today's world, the variety of options for communication, as well as cultural and linguistic diversity, demand a way of explaining how meaning that goes beyond written and oral language is constructed and is able to account for these other forms of visual, auditory, and dynamic representations (Kress, 2003; NLG, 1996). At the practical level and subsequent to this theoretical shift, a literacy pedagogy placed in specific local frameworks is needed (NLG, 1996) for promoting design in new technological environments, and for students and teachers to represent, interpret, and consciously manipulate language and information. This involves giving new meanings to literacy, and participating in socially valued activities (Dyson, 1997; Kalman, 2004; Myers, Hammet, and McKillop, 2000). Ideally, in any classroom, regardless of what subject is being taught, reading, writing, talking, and listening take place. In the particular case of a geography class, for instance, maps, images, and graphics are designed and interpreted, and it is necessary to support students and help them through these activities, and to know how to interpret and use these modes, so they can build a sense of what they read, listen to, and watch every day regarding the world in which they live (Roberts, 2003).

In this chapter I start from the assumption that it is possible to use technology to construct knowledge and learning (Jonassen, Peck and Wilson, 1999; Kress, 2003; Resnick, 2006). I therefore propose the notion of *technology in education* to be understood as the situated use of tools that contribute to students' learning.

Such use would be framed by the context as well as by different visions of learning, information, knowledge, curricula, and *affordances*[2] of technology. This involves moving beyond simple technological determinism to recognize ways in which the articulation of these elements can favor or restrict construction and negotiation of meanings, design, or the students' creative thinking.

Because it is a social practice, then, the use of technology is not neutral (Lankshear, Snyder, and Green, 2000; Street, 1984). Uses and benefits of technology are not universal as policies suggest; rather, they respond to ideological interests and are anchored to particular visions and local conditions of people who face implementation demands (Street, 2003). For these reasons, I have an interest in analyzing situated uses of technology and in portraying some of the power conditions particularly affecting academic communication practices (Lea and Street, 2006) that are also part of the context.

Policies and Ideologically Influenced Constructions of Technology

Middle school in Mexico is attended by adolescents who are between 12 and 15 years of age. This level of education became compulsory in Mexico only as late as 1993. In 2006 the Secretary of Public Education (SEP) began implementing an educational reform aimed at encouraging the students' capacities that would lead them to develop critical thinking and reflection, exercise civil rights and democratic participation, exchange knowledge through different means, be health conscious, protect the environment, and participate in an increasingly versatile labor market (DOF, 2006).

According to the SEP, information and communication technologies (ICT) "offer far reaching didactic and pedagogical possibilities" for educating the citizens of a "technologically advanced" society (DOF, 2006, 10) and for this reason one nodal element of this reform is the introduction of ICT. This tendency obeys, at least in part, the world wide pressures to reduce what the international policy agencies call the "digital divide":

> The division of the World into two groups: rich countries and poor countries, has become greater in the last 50 years. What is more surprising about this growing bimodal distribution of income is that it is not caused by differences in the accumulation of capital, but by differences in knowledge and in technology. The countries of the North have continued generating increasingly productive technologies, which according to statistical studies explain at least half of their economic growth. However, most countries of the South have not been capable to fully benefit from these new technologies; and, unfortunately, Latin America is inserted in the less dynamic group.
>
> *(Perry, De Ferranti, Gil, et al., 2003, 1).*

In this quote from the World Bank, countries are no longer seen as divided only geographically, economically, and politically, but now they are classified in terms of the "use" they make of new technologies. These tendencies have become concrete in the official Mexican documents emphasizing the presence of ICT and its supposed power for achieving educational changes.

> ICT favor interdisciplinary work in the classroom due to the fact that they offer the possibility to display multiple representations of a same situation or phenomenon on a screen.
>
> *(DOF, 2006, 10)*

It is taken as a fact that ICT on their own favor learning. Therefore, the role of the teacher in promoting interactions among students through the use of different resources becomes blurred. This is reflected in the economic investments that give priority to buying equipment over training teachers; and training remains a pending task (INEE, 2009). In this context, ICT acquire a *mythical* power (Guerrero, 2011) because of the extraordinary qualities attributed to them, independently of the fact that this may be true or not. The results obtained differ from this power (Cuban, 1986), because the use of ICT does not in itself guarantee improvement in teaching–learning processes (Kalman, 2006), or in students' achievements in the standardized examinations (Bruce and Hogan, 1998; Domaine, 2009). Information and communication technologies also do not themselves guarantee the yearned for equity in the use of technology (Warschauer, 2002).

However, the ideological exploitation of technology in education is similar to the assumptions applied to literacy in the 1960s, when it became the criterion that marked the difference between the civilized and the primitive (Goody and Watt, 1968) and was considered key for achieving economic development, democratization processes, and political participation (Kalman, 2008). The gaps assumed to be marked before by literacy are also now assumed because of technology use:

> just as ICT access is a prerequisite for full participation in the informational stage of capitalism, literacy was (and remains) a prerequisite for full participation in the earlier industrial stages of capitalism [...] they are both tied to somewhat controversial notions of societal divides: the great literacy divide and the digital divide.
>
> *(Warschauer, 2002, 6)*

To contribute to the discussion of what in fact constitutes those gaps, I will analyze classroom conditions and technology use by one teacher. In the same way that literacy does not only refer to skills and competences but also to social participation, the use of technology does not depend only on the presence of a machine, but on the social practices that are promoted through such use.

The Complex Context of the Classroom and Teaching Practice: Time, Spaces, Actors, Interaction, and Feelings within a Seventh Grade Classroom

The complexity of what happens at school is reflected in the relationship between stability and change (Cuban, 1986). Stability is present in certain rituals and norms (Jackson, 1975) and in elements of curricular structures such as contents, times, spaces, and resources (De Ibarrola, 2006). However, not everything that happens is predetermined, because the classrooms are also "cultures in the making," spaces in which members construct and reconstruct the events of everyday life. With their discourse and through interaction, they also construct and reconstruct social meanings (Collins and Green, 1992; Green and Dixon, 1993; Hicks, 1996; Yeager, 2006). Modifications in the classrooms can be detected through the details and changes in the interaction and collaboration among individuals (Sicilia and Lytras, 2005).

Given the impossibility of addressing all the dimensions in the classroom, I will only refer to some that came up through observation. These dimensions are: time, spaces, actors, interaction, and feelings within the classroom (Spradley, 1980).

The geography curriculum for 2006 in Mexico was intended to solve the problems that the 1993 curriculum experienced, such as the large amount of content that favored description, and made deep understanding and establishment of relationships between academic contents and the students' everyday lives difficult to achieve (SEP, 2006a). The current curriculum (SEP, 2006b) is made up of approximately 166 curricular topics that need to be covered in 175 classes of 50 minutes. However, much of that time goes to taking roll, getting students settled, and picking up the previous class's finishing place. A result of this is that effective teaching and learning time is reduced to approximately 35 minutes.

The strategy that Nico uses to cover the long list of topics is the following: first, he makes a presentation on the blackboard of the corresponding subject, and then this presentation is followed by questions and answers (Cuban, 1986; Mercer, 1997; see also Rockwell, this volume). On some occasions he leaves time for the students to voice their questions or to exchange comments. Sometimes Nico organizes the work in the classroom through groups; however, often while students work together, the teacher tends to hurry the students: "fast," "you have three minutes to finish." In one of these sessions the time was very limited to conclude the presentations of all the work that the students had produced within their teams during the class. Nico tried to allow all of them to present their results rapidly; therefore, he let some inaccurate assertions pass: for instance, some students confused cities with states and Nico did not correct them.

Some of the other teachers observed during data collection of the broader project use technology to meet the curricular content demand: they choose a video suggested by the Secretary of Public Education (SEP) that follows the order of topics as presented in the curriculum and, in this way, they cover several required contents in a brief amount of time; i.e. they cover a topic estimated to take up to 2 weeks in just 50 minutes, making time for covering other aspects of the curriculum. As a result, technology becomes a vehicle for transmitting information. Domaine (2009) calls this the instruction plug-in model, which reflects an epistemology in which knowledge is acquired by connecting the brain of the students to devices that allow them to receive the information and to use it. In other words, the assumption is that we learn *from* the technology, but not by doing something *with* it (Jonassen, Peck, and Wilson, 1999). At the time they select a video, the teachers consider that they "covered" the long list of required topics, and at the same time have complied with the demand of using ICT. What actually happened is that they did not modify the teaching patterns learned over years of professional practice.

Besides a lack of time, Nico also must meet the demands related to the use of spaces at school. The teacher goes daily to the classroom to meet with his class. When he decides to develop an activity involving technology he goes to another room called the Media Room. Since the 1990s, these types of rooms have been a common feature in most public middle schools: setting them up was an important part of the policy initiatives to make technology available to teachers and students. These rooms have a projector, a screen (often improvised by using a board painted white), televisions, a collection of videos, and between 20 and 30 computers that do not always work and whose repair represents a long bureaucratic procedure (Kalman, 2006). These elements vary according to the school. In some schools the use of the Internet is forbidden and in others it is allowed. In Nico's school the Internet is used under the supervision of the person in charge of the Media Room, Alex. During the field work it became apparent to me that these two teachers had a good working relationship, their interactions were informal, they joked often,[3] and generally seemed at ease. This relationship influenced the development of the students' activities involving technology, because, apart from positioning himself as an apprentice and collaborator, Nico felt confident enough to ask Alex to help him with an activity involving technology (Hicks, 1996; Tuyay, 1999). He also encouraged Alex's intervention in the class in which Nico presented the *PowerPoint®* presentation that both teachers had made together, to which I will refer further. This, the shared teaching practice between two teachers, is something that is not seen frequently in Mexican classrooms.

The possibility of change appears within the dimension of the interaction between all the participants; not only interactions between teachers but also between the teacher and students and through the feelings generated by these relationships (Spradley, 1980). Despite the fact that at times Nico appears authoritarian and extremely attached to the curriculum, he often promotes participation by his students: he allows them to go to the blackboard to explain something, or encourages them to ask questions, to make comments, or to express their feelings. Nico fluctuates between the heterogeneous teaching practices of a traditional or authoritarian teacher, and those of a flexible teacher (Guerrero, 2011). This is also reflected in the different visions he has about learning: sometimes he believes that learning is based on transmitting information; and other times (like the day he designed a lesson in which he made origami with his students) he seems to believe that learning is related to collaborating, and to helping his students achieve specific goals either with the help of others or by themselves (Vygotsky, 1995; Edwards and Mercer, 1987). In this origami class, Nico encouraged his students to build a spaceship. He demonstrated what he wanted them to achieve, accompanied them throughout the process, gave them time and opportunities to make mistakes, and promoted mutual help among them.

In the above description, I have depicted some of the aspects of schooling that illustrate the complexity of the classroom context and of teaching. While they cannot necessarily be generalized to all teachers, they illustrate elements of the reality of school that make it important for teachers and researchers alike to consider the uses given to technology as situated practices. Given the above, it can be concluded that it is the classroom practices, the way teacher and students organize and engage in activities that use computers and the Internet (and not the technology per se), that provide the possibility of bringing about authentic social learning and participation for students and teachers alike, as has been shown in research elsewhere (Cuban, 1986; Cuban, Kirkpatrick, and Peck, 2001; Lankshear, Snyder, and Green, 2000).

Nico Uses Technology for Teaching

The first time Nico tried to develop an activity with digital technology he decided to make a *PowerPoint®* presentation choosing the Victoria Falls as an example in order to talk about the Cultural Heritage of Humankind, a topic in the curriculum. Nico did not know how to use the software, so he asked Alex (the teacher in charge of the Media Room) for help. Alex showed him how to download images from the Internet and how to insert them in the presentation—in other words, he showed Nico how to operate the technological resource:

> Then we went on line with the teacher, eh: I came and told him what the topic was and we started searching … He, the teacher [Alex] helped me. I found the sources and then only started telling him what I needed […] (We are just starting to work like this) everything is new.
>
> *(N1290507)*[4]

Nico employs the plural to refer to the interaction between Alex and himself—"we got into," "we started," "we found" ("*nos metimos*," "*empezamos*," "*encontramos*") are phrases used to denote collaboration. The teachers designed a new type of text to be presented on screen, and even if there was little authorship in doing this, because they copied and pasted textual information from the Internet, the building process of the document is unquestionably social (Hicks, 1996). In the context of Mexican middle schools, this interaction is very significant, because collaboration among teachers is infrequent despite the initiatives of the SEP that have intended to promote collegiate bodies. These attempts for the most part have failed: teachers "learn to be alone within a collective, to individually sort out the issues of their work, school and professional lives on their own" (Sandoval, 2000, 345). Interaction among teachers reveals a new way to promote learning. Nico felt confident about asking Alex for help and positioning himself as learner and collaborator (Hicks, 1996; Tuyay, 1999).

The teacher took his students to the Media Room. Ten minutes of class time was spent on the students going upstairs to the second floor, forming a line in front of the room, and sitting in pairs. The physical distance between both classrooms is evident. The two of them are far apart. Because of these space issues, teachers who decide to go to the Media Room invest more time than necessary in activities involving technology, because they have to move from one place to another. The distribution model for computers at schools is based on the one for laboratories: this means that computers are placed facing the wall in special rooms that are located away from the conventional classrooms. This model seems to restrict the teachers' performance apart from prioritizing the operating management of ICT over other curricular learning (Jara, 2008) because, first, the teachers need to plan a special class; then, they need to organize their group to move to the Media Room; they need to figure out a way to sit the students before the computers; and they need to have projection or working materials ready. Other distribution models, like the ones that place some computers directly in the classrooms, seem to facilitate their use in a more fluid way during class time, because such computer use is promoted according to the development needs of the class: for instance, class work requiring looking up specific queries or solving particular questions.

Alex and Nico projected the presentation to the students. Alex instructed the students to open the file containing the same presentation located on the desktop of each computer. Meanwhile, Nico tried to help them so they could follow Alex's instructions on their respective computers. Nico anticipated only one of the *affordances*, that is, the possibilities of use offered by *PowerPoint*® in view of its social and physical characteristics (Kress, 2003; Kress and Bezemer, 2009): in this case, its use to display information. The selection of the resource is shaped by their social beliefs regarding its use (projection and transmission), the work of the teacher, and learning (Godwin and Sutherland, 2004), and many times these beliefs limit interactions within the classroom.

Nico started his presentation by showing the slides, and then he told the students: "You are going to … we are all going at the same time, eh?" (N1290507), as if he was anticipating that it would take his students less time to explore the presentation than what it would take him to actually present it. During the display, the teacher assigned turns to the students to read out loud, to repeat what was written on the slides. For Nico, the presentation had to be read and listened to by all at the same time. It is important to note that the repetition of information is a practice that the 2006 reform explicitly suggests modifying. The reform document notes: "The teaching practices that prevail in the classroom are: mechanical repetition of information and strong attachment to the textbook" (SEP, 2006a, 12). However, practices like the one observed in the Media Room were not documented in Nico's conventional classroom because there, before he starts to present a topic, he generally devotes a few minutes of class time to the students reading their textbooks individually and then presenting the material. During his presentation, students ask questions or make spontaneous contributions.

Nico tried to recover this way of interacting with his students by promoting their participation during the *PowerPoint*® presentation and when he tried to pose questions about the measures being taken to protect the site that was included in his slides. He asked things like: "What do you think about this?" "What else can you comment about that?" "What is the intention of this message that was given to us?" These questions, apart from resulting in ambiguity for the students, reflect the way Nico distanced himself from the information presented in the paragraph drafted by unknown authors (i.e. when he says: "given to us"); regardless of the fact that he had made the presentation, he showed limited ownership of what was written in it.

The material character of the object from where the students read (in this case the computer screen) is important, because during the reading protocols it is possible to give plural meanings to the texts (Chartier quoted by Rockwell, 2001) despite the fact that their material character is the same, people can construct different versions of the same object, generate interactions with other community members and gain learning experiences from them (Pahl and Rowsell, 2010). In this context the computer was used to confirm a way of reading that restricts the opportunities of appropriation and construction of meaning as well as to promote very precise interactions between teacher and students because this way of reading is based on the mechanical act of repetition of the information displayed. This reading protocol was favored by the display of the slides' presentation on the screen. The fact that the teachers have set ideas about the restrictions or possibilities of technology, and that they articulate them with their ideas about school activities (in this case reading), can sabotage or contribute in a negative way to the communication goals within the classroom. In this example, the *PowerPoint*® presentation worked as a *plug-in* resource, namely, to present or transmit information, and to restrict conversation instead of promoting it (Domaine, 2009).

After finishing reading out loud from the slides with text, Nico asked the students to explore the slides with images: "And as from now, you will only look at all the slides that we have, freely [...] free access, until the map" (N1290507). The expression "free access" means that the teacher authorized the students to advance the slides at their own pace. The students explored spontaneously, without receiving suggestions regarding where to focus their attention. Most of them finished the task in under four minutes; two of them in seven seconds. Nico asked the students who had finished looking at the slides to raise their hands. When he noticed that most of the students had finished looking at them in very little time, he added: "But calmly, ah? Because if you pass them too quickly, it doesn't work" (N1290507).

The teacher had put emphasis on the slide with the map because it allowed him to legitimize the geographic content that was included in the ICT activity. However, when he got to the map, the teacher limited his comments to a superficial description of some of its features:

> P: Well, I believe that most of us are already in the map. Those who are in the map, they mark for us there the riverbed of the river from the entire zone that we are looking at, and especially they are framing for us the Victoria cataracts. There, and then, they mark for us two air routes of people who make the tours for this place; we are speaking about the river Zwambeze in the central part of Africa. The following one please [...] Free access.
>
> *(N1290507)*

Nico is the one who describes the main features of the map. He seems to believe in a possible beneficial effect of images in the learning process of his students; that is, he has the idea that, by merely looking at the images, his students will acquire knowledge; as if the meaning of the images would depend only on the image itself, and not on the interpretative activity of the student reader. Though objects can have certain meaning in terms of their material character, for some people and in certain areas they may have no value. However, to the extent in which teachers promote further possibilities of interpretation and construction of meaning derived from them, these objects can promote powerful learning (Pahl and Rowsell, 2010). This activity with technology, in difference from other activities such as that of making origami, was not one in which the students showed themselves intellectually committed in terms of what the 2006 reform demands or expects.

When the exploring of slides finished, the teacher gave instructions for homework: "Now, kids, what we want is more or less a situation like this in your own work ... to be precise, you need to look up several Internet pages so you can get information, specially about the slides, because you need to look for the sites where all those pictures of places come from" (N1290507).

Nico asks the students to look for Internet pages, to extract information, to find the pictures, and to make slides. He is asking them to reproduce the exact procedure followed by both teachers to make the presentation; a procedure that they did not make explicit to the students because the students only saw the final product. This aspect differs from what Nico does in other classes where he accompanies the students throughout the development of their assignments.

What this Episode Suggests

The classroom is a complex environment with multiple interrelated dimensions that, along with the social interactions that occur there, give shape to the context where technology insertion takes place. Stability is perceived by the large amount of curricular contents and little time to cover them, or in the concepts of learning that equate it with the transmission of information. Nico's case shows how the use of technology requires a different configuration of stable elements, i.e. time and space.

Even though using technology offers the possibility to construct knowledge, in the case analyzed it appears as a straitjacket for the teacher's actions. In other circumstances he shows teaching practices that are more flexible. The teacher uses the software to transmit information, and by doing this he restricts the students' opportunities to make original contributions. This does not happen when he is not using technology.

We also observe a limited understanding of the *affordances* offered by technology (apart from the projection of information). The teacher asked the students for a similar product to the one he presented, but he left the gradual process required to build this product in the dark. This differs from other classes (e.g. origami) in which he accompanies his students and demonstrates the activity. Nevertheless, the teachers produced an emergent text in a social way (Hicks, 1996); a product made in and for the screen. This highlights the need to "add value" to the technical uses made by the teachers and to propose new professional development processes to bring them closer to the uses of technology for producing and representing meanings (Lankshear and Knobel, 1997).

The use of technology represents an opportunity to break with the teacher's isolated practices that frequently, although not always, occurred in usual classroom practice. Collaboration between this teacher and the one in charge of the Media Room, even though it was related to operating aspects, shows the importance of mediators (Kalman, 2003) in constructing knowledge, and the need for teachers to move away from a role in which they are information providers to another one in which they can be apprentices and collaborators (Tuyay, 1999). However, the pre-established trust among the teachers influenced this collaboration; therefore one cannot think it possible to simply trigger this trust by decree.

Conclusion: Technology Situated in New Contextual Configurations

Throughout the chapter I have documented the situated nature of the use of technology in schools. My interest was to analyze the local meaning of this practice and to avoid simplistic assertions (Street, 2009) related to literacy pedagogies that incorporate the use of technologies.

There is a mythic and idealized construction of technology. However, what happens in the classroom clearly shows how its presence does not bring about changes by itself, but that these are favored or restricted by the conditions and the context. It is, then, not possible to think about introducing technology or training teachers outside their concrete contexts. To speak about technology in the classroom implies thinking about it in terms of situated practices and social uses (and not only as distribution of equipment), in terms of the interactions that need to be promoted between teachers and students, and in terms of the necessary curricular adjustments to promote the use of technology to construct learning. In order to propel these changes, it is necessary to think of new ways of promoting collaboration as well as authentic interactions among teachers; to create new ways of conceiving the curriculum—much more powerful than those that refer only to addressing contents in a superficial and orderly way; to analyze the current notions of learning and technology that occur within the classroom; and to re-imagine all these notions from the perspective of the teaching practice. This is because what might cause powerful changes in the teaching practices are the arrangement and the weight of the social and technological elements, and not technology by itself.

Understanding these possible configurations could be suggestive for creating new learning opportunities through the use of ICT and for making evident the ones that already exist in the classroom. As shown by the examples above, it is not possible to expect a pedagogical or a technological revolution by decree, but rather a succession of very gradual changes that in the very long term will end by transforming the educational scenario (Cuban, Kirkpatrick, and Peck, 2001).

Notes

NOTE: The author of this chapter has used pseudonyms throughout to protect the identities of students and instructors.

1 Doctoral dissertation "Ahí está el detalle: Cambios minúsculos, rutas opacas y tecnologías míticas en la enseñanza de la geografía en secundaria" ("That is the point: small changes, blurred roads and mythic technologies in the teaching of geography in middle school") that I wrote under the guidance of Dr. Judith Kalman at the Department of Educational Research of the Research Center for Advanced Studies. (Departamento de Investigaciones Educativas [DIE] del Centro de Investigación y Estudios Avanzados [Cinvestav]. The research was financed by the National Council for Science and Technology (Consejo Nacional para la Ciencia y la Tecnología [Conacyt] and inserted in the research agenda of the Laboratory of Education, Technology and Society (Laboratorio de Educación, Tecnología y Sociedad).

2 *Affordances* (Kress, 2003) refers to the possibilities and limitations that a socially and culturally configured mode or "way" of representing knowledge (image, audio, writing, distribution of a page) offers. It allows a meaning maker to analyze the potentials of a "way" or a given semiotic resource and to (1) question its possibilities and limitations; (2) discern what it facilitates or not, in comparison with other "ways"; and (3) appreciate the achievement of different types of semiotic work and certain aspects of significance in his (or her) message. Said analysis is carried out in relation to his (or her) interests and characteristics of the audience to which this message will be destined (Kress and Bezemer, 2009).

3 For example, during Alex's visit to Nico's classroom, Alex asked him if he had obtained the music of a popular singer. In this communicative situation of collusion, both

laughed expressing agreement (McDermott and Tylbor, 1995). They also joked about the condition of the computer that did not work and was placed at the classroom in which Nico was teaching.

4 The codification employed in the transcription was: lengthen syllables::; pauses…; better approximation to what is inaudible ( ); comments related to actions (); suppression of fragments […].

References

Bruce, B., and Hogan, M. (1998). The disappearance of technology: toward an ecological model of literacy. In D. Reinking, M. McKenna, L. Labbo, and R. Kieffer (eds.), *Handbook of Literacy and Technology*. New Jersey: Lawrence Erlbaum Associates, Inc., 296–311.

Cazden, C. (1988). *Classroom Discourse: The Language of Teaching and Learning*. Portsmouth, NH: Heinemann.

Coates, J. (1996). *Talk: Conversation between Woman Friends*. Cambridge, Massachusetts: Blackwell Publishers Inc.

Collins, E., and Green, J. (1992). Learning in classroom settings: making or breaking a culture. In H. Marshall (ed.), *Redefining Student Learning*. Norwood, NJ: Ablex, 59–85.

Cuban, L. (1986). *Teachers and Machines: The Classroom Use of Technology Since 1920*. New York: Teachers College Press.

Cuban, L., Kirkpatrick, H., and Peck, C. (2001). High access and low use of technologies in high school classrooms: explaining an apparent paradox. *American Educational Research Journal* 38 (4), 813–34.

De Ibarrola, M. (2006). La formación para el trabajo en las escuelas: el currículo y las prácticas formativas [Training for work at school: curriculum and formative practices]. In M. De Ibarrola, *Formación escolar para el trabajo: posibilidades y límites. Experiencias y enseñanzas del caso mexicano* [*Training for Work at School: Possibilities and Limitations of the Mexican Experience*]. Montevideo: Cinterfor, 151–210.

Diario Oficial de la Federación [DOF] (2006). *Acuerdo número 348 por el que se establece el nuevo Plan y Programas de Estudio para Educación Secundaria*. México, D.F., Official Journal of the Federation (2006). [Agreement No. 348 by which the new Plans and Programs for middle school were established.] Mexico, D. F. http://www.consejosconsultivos.sep.gob.mx/P%C3%A1gina%20Web/plantillas/documentos/acuerdo_384.html (February 21, 2008).

Domaine, V. (2009). *Rethinking Technology in Schools PRIMER*. New York: Peter Lang.

Dyson, A. (1997). *Writing Superheroes: Contemporary Childhood, Popular Culture, and Classroom Literacy*. New York: Teachers College Press.

Edwards, D., and Mercer, N. (1987). *El conocimiento compartido: el desarrollo de la comprensión en el aula* [*Common Knowledge: The Development of Understanding in the Classroom*], R. Alonso Pérez, trans. Barcelona: Ministerio de Educación y Ciencia and Paidós.

Erickson, F. (2004). *Talk and Social Theory*. Malden, MD: Polity Press.

Ferreiro, E. (2010). Ceremonia de nombramiento como Investigadora Emérita del Cinvestav. México D.F., 3 de diciembre, 2010. [Appointment Ceremony as CINVESTAV Emeritus Researcher]. Mexico City, December 3, 2010.

Floriani, A. (1994). Negotiating what counts: roles and relationships, texts and contexts, content and meaning. *Linguistics and Education* 5, 241–74.

Godwin, S., and Sutherland, R. (2004). Whole-class Technology for Learning Mathematics: the case of functions and graphs. *Education, Communication and Information* 4 (1), 131–52.

Goody, J., and Watt, I. (1968). The consequences of literacy. In J. Goody (ed.), *Literacy in Traditional Societies*. Cambridge: Cambridge University Press, 1–68.

Green, J., and Dixon, C. (1993). Talking knowledge into being: discursive and social practices in classrooms. *Linguistics and Education* 5, 231–9.

Green, J., and Bloome, D. (1997). Ethnography and ethnographers of and in education: a situated perspective. In J. Flood., D. Lapp., and S. B. Heath (eds.), *Handbook for Literacy Educators: Research in the Community and Visual Arts*. New York: Macmillan, 181–203.

Guerrero, I. (2011). *Ahí está el detalle: cambios minúsculos, rutas opacas y tecnologías míticas en la enseñanza de la Geografía en secundaria*. [That is the point: small changes, blurred roads and mythic technologies in the teaching of geography in middle school]. Unpublished doctoral dissertation, Mexico: DIE-Cinvestav.

Gumperz, J. (1999). Sociocultural knowledge in conversational inference. In A. Jaworski and N. Copland (eds.), *The Discourse Reader*. London: Routledge, 98–106

Heath, S., and Street, B. (2008). *On Ethnography: Approaches to Language and Literacy Research*. New York: Teachers College Press.

Hicks, D. (1996). *Discourse, Learning, and Schooling*. Cambridge: Cambridge University Press.

Instituto Nacional para la Evaluación de la Educación (2009). *Panorama Educativo de México 2009 Indicadores del Sistema Educativo Nacional. Educación básica* [Mexican Educational Outlook 2009. National Basic Education System Indicators]. Mexico City: Instituto Nacional para la Evaluación Educativa.

Jackson, P. (1975). *La vida en las aulas* [*Life within the classrooms*], A. Gutierrez, trans. Madrid: Morova.

Jara, I. (2008). *Las políticas de tecnología para escuelas en América Latina y el mundo: visiones y lecciones* [*Technology Policies for Schools in Latin America and the World: Visions and Lessons*]. Santiago de Chile: CEPAL, http://www.cepal.org/SocInfo (October 14, 2009).

Jonassen, D., Peck, K., and Wilson, B. (1999). *Learning with Technology: A Constructivist Perspective*. Ohio: Prentice Hall.

Kalman, J. (2003). *Escribir en la plaza [Writing on the Plaza]*. Mexico City: Fondo de Cultura Económica.

Kalman, J. (2004). *Saber lo que es la letra: una experiencia de lecto-escritura con mujeres en Mixquic, México* [*Discovering Literacy: Access Routes to Written Culture for a Group of Women in Mixquic, Mexico*]. *México, D.F.*: Secretaría de Educación Pública-UIE-Siglo XXI Editores.

Kalman, J. (2005). Mothers to daughters, *pueblo to ciudad*: women's identity shifts in the construction of a literate self. In Alan Rogers (ed.), *Urban Literacy: Communication, Identity and Learning in Development Contexts*. Frankfurt: Unesco Institute of Education, 83–103.

Kalman, J. (2006). *Incorporación de la Tecnología de la Información y la Comunicación (TIC) a la práctica docente en la educación secundaria: Informe presentado a la Secretaría de Educación Pública* [*Incorporating Information and Communication Technology (ICT) to Teaching Practices in Middle Schools: Report Presented to the Secretary of Public Education. Mexico*]. Unpublished manuscript, CINVESTAV-ILCE. Mexico City.

Kalman, J. (2008). Literacy in Latin America. In B. Street and N. Hornberger (eds.), *The Encyclopedia of Language and Education*. New York: Springer, 321–34.

Kalman, J. (2009). "San Antonio: ¡Me urge!" Preguntas sin respuesta acerca de la especificidad de dominio de los géneros textuales y las prácticas letradas ["Saint Anthony, it's urgent!" Unanswered questions about the domain specificity of textual genres and writing practices]. In J. Kalman and B. Street (eds.), *Lectura, escritura y matemáticas como prácticas sociales: Diálogos con América Latina* [*Reading, writing and mathematics as social practices: Dialogues with Latin America*]. Mexico City: Siglo XXI Editores and CREFAL, 130–55

Kress, G. (2003). *Literacy in the New Media Age*. London: Routledge.

Kress, G., and Bezemer, J. (2009). Escribir en un mundo de representación multimodal [Writing in a world of multimodal representation]. In J. Kalman and B. Street (eds.), *Lectura, escritura y matemáticas como prácticas sociales: Diálogos con América Latina* [*Reading, writing and mathematics as social practices: Dialogues with Latin America*]. Mexico City: Siglo XXI Editores and CREFAL, 64–83.

Lankshear, C., and Knobel, M. (1997). Literacies, texts and difference in the electronic age. In C. Lankshear, *Changing Literacies*. Buckingham: Open University Press, 132–63.

Lankshear, C., Snyder, I., and Green, B. (2000). *Teachers and Technoliteracy: Managing, Literacy, Technology and Learning in Schools*. Australia: Allen & Unwin.

Lea, M., and Street, B. (2006). The "Academic Literacies" Model: Theory and Applications. *Theory into Practice* 45 (4), 368–77.

McDermott, R., and Tylbor, H. (1995). On the necessity of collusion in conversation. In D. Tedlock and B. Mannheim (eds.), *The Dialogic Emergence of Culture*. Urbana, IL: University of Illinois Press, 218–36.

Mercer, N. (1997) *La construcción guiada del conocimiento: el habla de profesores y alumnus* [*The Guided Construction of Knowledge Construction: Talk Amongst Teachers and Learners*], Inésde Gispert, trans. Barcelona: Paidós.

Myers, J., Hammett, R., and McKillop, A. (2000). Connecting, exploring and exposing the self in hypermedia projects. In M. Gallego and S. Hollingsworth (eds.), *What Counts as Literacy: Challenging the School Standard*. New York: Teachers College Press, 85–105.

New London Group (NLG) (1996). A pedagogy of multiliteracies: designing social futures. *Harvard Educational Review* 66 (1), 60–92.

Pahl, K., and Rowsell, J. (2010). *Artifactual Literacy: Every Object Tells a Story*. New York: Teachers College Press.

Perry, G., De Ferranti, D., Gill, I., Guasch, J., Maloney, W., Sánchez-Páramo, C., and Schady, N. (2003). Cerrando la brecha en educación y tecnología: Estudio del Banco Mundial sobre América Latina y el Caribe [Closing the gap in education and technology: a World Bank report on Latin America and the Caribbean], 25, 1–4.

Resnick, M. (2006). Computer as Paintbrush: Technology, Play, and the Creative Society. In Singer, D., Golinoff, R., and Hirsh-Pasek, K. (eds.), *Play = Learning: How play motivates and enhanced children's cognitive and social-emotional growth*, Oxford University Press, 192–208.

Roberts, M. (2003). *Learning through Enquiry*. Sheffield: Geographical Association.

Rockwell, E. (2001). La lectura como práctica cultural: conceptos para el estudio de los libros escolares [Reading as a cultural practice: concepts for the study of text books]. *Educação e Pesquisa* [*Education and Research*], 27 (1), 11–26.

Sandoval, E., (2000). *La trama de la escuela secundaria: instituciones, relaciones y saberes* [*The plot of the middle school: institutions, relationships and knowledge*]. Mexico City: Universidad Pedágogica Nacional & Plaza y Valdés.

SEP (2006a). *Reforma de la Educación Secundaria. Fundamentación curricular: Geografía de México y del Mundo* [*Middle School Reform: Curriculum Fundamentals—Mexican and the World Geography*]. Mexico City: Dirección General de Desarrollo Curricular de la Secretaría de Educación Pública (DGDC-SEP).

SEP (2006b). *Educación básica. Secundaria. Geografía de México y del Mundo. Programa de estudio 2006* [*Basic education: middle school. Curriculum fundamentals. Mexican and the World Geography*]. Mexico City: DGDC-SEP.

Sicilia, M., and Lytras, M. (2005). On the representation of change according to different ontologies of learning. *International Journal of Learning and Change* 1 (1), 66–79.
Spradley, J. (1980). *Participant Observation*. New York: Holt, Rinehart and Winston.
Street, B. (1984). *Literacy in Theory and Practice*. Cambridge: Cambridge University Press.
Street, B. (2003). What's "New" in New Literacy Studies? Critical approaches to literacy in theory and practice. *Current Issues in Comparative Education* 5, 1–14.
Street, B. (2008). New literacies, new times: developments in literacy studies. In B. Street and N. Hornberger (eds.), *Encyclopedia of Language and Education*. New York: Springer Science+Business Media LLC, 418–31.
Street, B. (2009). Perspectivas etnográficas y políticas sobre cultura escrita: el poder de nombrar y definir. [Ethnographic perspectives and policies about literacy: the power of naming and defining]. In J. Kalman and B. Street (eds.), *Lectura, escritura y matemáticas como prácticas sociales: Diálogos con América Latina* [*Reading, Writing and Mathematics as Social Practices: Dialogues with Latin America*]. Mexico City: Siglo XXI Editores and CREFAL.
Tuyay, S. (1999). Exploring the relationships between literate practices and opportunities to learn. *Primary Voices K-6: National Council of Teachers of English* 7, 17–24.
Vygotsky, L. S. (1995). *Pensamiento y lenguaje* [*Thought and Language*]. Barcelona: Paidós.
Warschauer, M. (2002). Reconceptualizing the digital divide. *First Monday* 7 (7), 16–18.
Yeager, B. (2006). Teacher researcher/researcher teacher: shifting angles of vision for making visible what students can do. *Language Arts Journal of Michigan* 22, 26–33.

13

PREAMBLES, QUESTIONS, AND COMMENTARIES: TEACHING GENRES AND THE ORAL MEDIATION OF LITERACY

Elsie Rockwell

Introduction

Meaning is not inherent in texts but rather created through social interaction. This is true even in schools where authorities attempt to stabilize curricular contents through the distribution of uniform textbooks, and thus to guarantee a common basic education. Students' relationship to school texts is largely mediated through surrounding interactions both in and out of the classroom. The teachers' oral discourse has particular import, as it models, guides, and evaluates students' relationship to textbook lessons, providing cues to ways of interpreting and producing meaning while reading the text.

The literacy practices discussed below involve the use of particular speech genres in elementary classrooms where textbooks were the central, and at times the sole, written reference to curricular contents. Educators in the latter part of the 20th century in Mexico often assumed that the official textbooks placed heavy constraints upon teaching and homogenized classroom representations of curricular contents. My ethnographic research has pointed rather to the multiple ways in which teachers mediate the use of textbook lessons, often reflecting diverse pedagogic traditions that converge at a given time (Rockwell, 1995, 2000, 2007; Guerrero, this volume). In this chapter, I compare discourse patterns of three quite different teachers during tasks that involve students in reading history textbook lessons in Mexican classrooms.[1] Through the use of particular speech genres, these teachers mediate the textbook lessons in different ways, orienting students' interpretations towards distinct meaning systems. By analyzing contextualization and intertextuality in the teachers' discourse, I suggest that very different ways of establishing the meanings of particular texts—as well as the meaning of text itself—are in play. In light of this analysis, I argue that schooled literacy practices do not produce a uniform "literate bias" in the transmission of academic content. Rather, the oral–written matrix of these classrooms reveals a range of literacy practices and of ways of verbally mediating students' relationship to texts.

Language and Literacy in the Classroom

I approach reading and writing by reconstructing literacy practices through the ethnographic study of events that involve the use of written text (Street, 1984). In this perspective, all uses of writing are seen as social practices (even solitary reading) in that they are constituted within and through social relations and are embedded in broader social contexts and processes (Barton and Hamilton, 2005). As social practices, classroom literacy practices are set within communities and networks, and are often framed within structures of formal schooling marked by asymmetrical power relations, where students and teachers negotiate everyday class work. From this perspective, a literacy event such as reading a particular lesson is not self-contained, as textbooks bear the marks of their processes of production in other contexts, and both teachers and students mobilize skills and meanings from a variety of contexts towards the interpretation of and work with texts (Barton and Hamilton, 2005).

I further consider literacy practices as cultural practices, a concept which adds different dimensions to the analysis. Scholars who regard uses of writing as cultural practices (e.g. Chartier, 1993) focus on the cultural tools and signs involved and on the ways literacy practices are recognized and appropriated—and in some cases named or renamed—within particular cultural communities or networks. In underscoring the cultural dimension, researchers situate literacy practices in shared structures of meaning and in historically generated chains of transmission and appropriation (Bakhtin, 1986). Particular practices are recognized, acknowledged, interpreted, and reproduced, albeit with constant modification, in appropriate social situations. An example of a classroom cultural practice involving text is choral or rote recitation of textbook lessons, a widespread and recognizable practice within past and present school cultures.

Approaching school literacy practices that are embedded in oral discourse opens the issue of the constraints that asymmetric power relations and specific participant structures place upon verbal forms used in classrooms. The classical discussion involves the weight of "triadic dialogue" (Lemke, 1990), also referred to as the IRE or IRF pattern (for Initiation-Response-Evaluation/Feedback), in interactions between teacher and student. This verbal form has been considered a default pattern of classroom discourse (Cazden, 1988/2001) and its prevalence has been documented in many contexts. However, triadic dialogue, sometimes considered "a sort of genre" (Lemke, 1990, 123), is certainly not the only genre used by teachers in their interaction with students. Studies have revealed both significant variations in the triadic pattern and an array of other verbal forms of interaction in classrooms, including distinctly marked speech genres ranging from sermons to signifying.

In a previous publication (Rockwell, 2000), I used evidence from discourse patterns of one teacher to show that "teaching genres" are an example of what Bakhtin (1986) called "secondary" or "complex" genres, in that they combine a variety of primary speech genres within the formal constraints of classroom interaction. Some genres, for example dictation, are considered specific to classroom instruction, and bear markers, such as intonation, that make them easy to recognize; students respond to the cues of dictation by taking up their pencils and writing certain parts of what the teacher says while responding differently to other parts (for example, they omit repetitions and respond appropriately to instructions such as "underline" or "in red"). An important aspect of the Bakhtinian concept of genre is that it is not limited to formal characteristics of discourse; it also refers to the substantive content represented in particular discursive forms. Thus, pieces of school knowledge become conventionalized (Edwards and Mercer, 1989) and encoded in generic forms, and are continually reproduced and subtly transformed in particular school cultures. For example, particular versions of Mexican history—such as the struggle for Independence led initially by Miguel Hidalgo—are recurrent in the classrooms and school ceremonies observed in this and other studies, and take on generic qualities, including formulaic phrases ("freedom from slavery", "rise up in arms") that are readily acknowledged by students.

Further discussion of speech genres enriches their use in describing classroom discourse. Bauman (1992, 132) stresses the dialogic nature of genres, in the Bakhtinian sense, "resting not primarily on dialogue in the turn-taking social interactional sense ... but on the interplay of two (or more) primary genres, each with its own formal and functional characteristics." Attention to the interplay of genres captures one aspect of intertextuality, the significant linkages between discourses that are juxtaposed within a single text or utterance (Bloome and Egan-Robertson, 1993; Briggs and Bauman, 1992). Bauman's approach to intertextuality focuses on contextualization, a process through which performers situate their discourse in contexts that lend them authority or render them persuasive, by establishing webs of intertextual linkages to particular cultural systems of meaning and interpretation (Bauman, 1992, 128, 142). Contextualization through intertextual reference is one of the ways teachers mediate school texts.

Given the prevalence of contextualization in classrooms, it is important to keep in mind an argument developed years ago by Jean Lave, against the view of schooling as a setting for "decontextualized" learning (Lave, 1996). Classrooms, as all other settings, are constructed as social contexts and sites for engaging in context-specific learning practices. Viewing classrooms as particular sorts of contexts forces us to ask questions regarding the complexity of activities involving children, teachers, textbooks, and meaning, set within ongoing social interaction. In what follows, I explore how texts are contextualized in classrooms through intertextual reference, and how in some cases they become seemingly "decontextualized" through the use of specific, context-laden speech genres.

Contextualization involves *linguistic mediation*, in the sense put forth by Baynham and Masing (2000, 207): "... the crossing/diminishing of distance linguistically, in terms of shifting to more familiar from more distant registers ..." By analyzing classroom mediation, I shift the focus from the students' direct relationship to the textbook towards their experience of "reading a lesson" as a literacy practice shaped by/with their teachers.[2] I select moments in which teachers engage in verbal interaction with students while parts of the texts are taken up in the oral performance of reading. During such sequences, the distinction between oral and written modes becomes blurred. The teachers' oral interventions encompass and draw attention to particular aspects of the texts and establish intertextual links with other discourses. They serve as benchmarks for interpretation of the texts and signal accepted modes of relating to textual knowledge in the classroom. By studying variation in speech genres used in mediation, we may gain insight into the different ways in which schooled literacies build the semiotic environments within which learning takes place.

The Cultural Context of Classroom Observations

The present analysis draws on ethnographic research in elementary schools in two municipalities, Contla and Chiautempan, in the state of Tlaxcala in central Mexico, during several periods of fieldwork (the 1980s, 1990s, and 2010–2011). Within the past 30 years paved roads, public transportation, electricity, bureaucracy, and schooling have extended the urban zone past the municipal centers into the once rural areas. In the smaller towns on the hillside, where observations were made, some families still maintain small cornfields of vegetable gardens and raise poultry or herd sheep, though everyday life takes them to the municipal head towns and nearby cities for purchases and medical attention, jobs in factories or services, and higher education. The region offers interesting evidence of an oral–written matrix over an array of domains, as literacy practices are interwoven and articulated with oral discourse in complex ways. Within this social context, as most others, an oral/literate dichotomy does not hold. As elsewhere, uses of writing are context-bound, and their significance is a consequence of local culture and social relations (Rockwell, 2010).

This is equally true for classroom contexts. Since 1960 the national Secretariat of Public Education has published and delivered free of charge to each elementary school child graded texts and workbooks covering all primary school subjects. Textbooks mobilize specific contents defined by distant actors—in this case the national teams involved in periodically producing the new curricula and writing textbooks, who in turn draw on current scholarship—and bring them into the classroom in a transportable form (Nespor, 1997). Though graded readers and a few texts were available before this program, the free textbooks produced significant changes, as they became the most tangible expression of the official curriculum within the classroom, and provided students with textual and graphic representations of contents that were often different from those offered by teachers. However, the textbooks did not determine uniform teaching practices or learning outcomes; rather, the uses of textbooks in class were variously defined by teaching traditions ranging from the most ritualized modes of recitation to more open forms of teaching. Texts were received and interpreted by both teachers and students within preexisting structures of meaning.

The following three examples include very brief extracts taken from observation protocols and audio-recordings of classroom interaction. In all cases, the lessons included reading parts of the history text; however, the practices varied from individual students reading aloud to the class, to silent reading by all students, to the teacher reading the text aloud while students followed in their own textbooks. In all three cases, teachers also posed comprehension questions, though the manner differed: the first teacher asked students to answer questions orally, while writing the answers on the board; the second involved dictated questions to be written and answered individually by the students as deskwork; the third incorporated oral questions into her own ongoing commentary on the text. I examine the generic texture of each extract in turn as well as the intertextual references and contextualization implied in the teachers' oral mediation of the texts.

A Preamble to the Text: The Idea of Liberty

The first class, a fifth grade observed in 1981, was taught by Eufemio,[3] a teacher with 30 years of experience, trained in the normal schools of the post-revolutionary period. He was principal of a small school in the municipal head town, and was substituting for one of his teachers, who was absent that day. This textbook lesson was on "the idea of liberty" that began to change the political panorama in Europe, and began as follows:

> THE IDEA OF LIBERTY[4]
> *All human groups need a government and laws in order to live and work in peace. As time passes, human needs, ideas, customs and interests change, and so the forms of government and laws change. For this reason, we find different types of government, and different ideas and opinions on the best way to govern a country.*
>
> *We will study several examples.*
>
> *In Europe, during the Middle Ages, social organization was unjust: there were nobles who possessed lands and serfs who worked for the nobles ...*[5] *[Continues on the Middle Ages and rise of cities.]*

Eufemio asked the children to open their history textbooks to this lesson. He then read the title and immediately offered them what he later termed a "preamble."

During his intervention, lasting several minutes, he anticipated the content of the lesson and expanded the notion of the "idea of liberty" with reference to topics more familiar to the students. The tone and intertextual connections of this preamble can be observed in the following excerpt:

> T: (he holds the textbook open in front of the class, students have their own texts as well)[6] [.....] we have not always had liberty... because there has not always existed the same type of government ... [.....] once there were kings ...monarchs ... and slavery [interruption at the door] we have said that the world community at different times has suffered terrible slavery ... and those people who suffered always dreamt with ideas of liberty [.....] as I read the idea of liberty ... I imagine the peoples that were enslaved in antiquity [.....] for example, in Mexico ... Europeans had their colonies in Mexico ... in Africa [.....] those Europeans who arrived with more culture and more technology tried to take over land here to establish their industry ... and earn capital ... but what was the situation (?) ... they came to exploit cheap labor ... to install industry and commerce [.....] then the people, anxious, desiring liberty ... began to have the idea of liberty ... that Mexico should not be dominated by Spain [.....] do you remember who freed us from slavery (?)... (students answer Miguel Hidalgo) [.....] so... that is in general what you are going to study ... later you will see it in detail... [.....] this is the preamble ... now we are going to read

Eufemio presented the content of the lesson within a discourse characteristic of the tradition of teaching associated with the post-revolutionary period. The content of the textbook lesson refers to social changes occurring after the Middle Ages, as ideas of liberty spread in European nations; it mentions serfdom as a characteristic of the feudal age and attributes the struggle for freedom to the dissemination of the "idea of liberty." The preamble, by contrast, includes intertextual references to an encoded version of social injustice and the evolution of societal forms, from feudalism to capitalism, including the establishment of colonies and exploitation of labor by Europeans in the New World. This history was disseminated in Mexico during the years of socialist education (1930s) and survived among the elder teachers of the region. The teacher further contextualized the lesson by connecting serfdom with slavery, a term easily associated with the well-known episode in the struggle for Mexican Independence from Spain, when Hidalgo declared independence and proclaimed the abolishment of slavery (not accomplished until years later) in the Spanish colonies. Independence Day speeches commonly state that Hidalgo "freed us [all] from slavery," extending the notion of slavery to the colonial position of the nation as "subjected to Spain," or "under the yoke of Spanish rule." This version constitutes a recognizable speech genre, a conventional rendering of national history.

As students took turns reading the text, Eufemio continued to comment and inserted rhetorical questions, which the students had little difficulty in answering as he gave sufficient cues to elicit rote "yes" or "no" answers. Though framed in the IRE/F pattern, these questions were not testing students' previous knowledge; rather the teacher sought verbal acquiescence with the meaning of his preamble. Several questions verged on the "injustice" of the feudal lord/serf relations, and at times linked the content of the text to the present-day experience of the children, a discursive move that was common among teachers of his generation. In some of these exchanges students seemed unwilling to answer. For example:

1. T: as human beings, do we not have the right to live in equality?
2. Ss: (in unison) Yes ...
3. T: it was as though a feudal lord were here, and told you, 'come and work my fields', and you harvested his crops and gave them to him, and then he would give you (of the crops) only the indispensable ... that was the Middle Ages, we were all in darkness [...] was it fair that only a few people owned land and the rest were workers(?)
4. Ss: no (only a few students answer no, though cues contained in the structure of the question that calls for a negative seem to be sufficient).
5. T: now, though, can't we all own land? (Most students remain silent, two or three say yes, in a low voice).
6. T: yes, we all can ... some have more plots, because they have bought them, but other who do not, can buy land. In that time only the nobles could have land.

The teacher attempted to situate the students in relation to the text in turn 3: "It was as though a feudal lord were here, and told you, 'Come and work my fields ...'" This re-creation of a hypothetical situation involving the students, strengthened by the use of direct discourse, is a traditional teaching genre found in older pedagogical dissertations. The teacher continued by establishing a contrast, stating that in feudal times only nobles could have land, while in the present day all could own land.[7]

The nature of the questions posed both during the initial preamble and while reading the text related the contents to meanings or terms that were more familiar to the students, both through previous schooling (the story of Hidalgo and the abolition of slavery) and in their everyday lives (the ownership of land). It also allowed the teacher to introduce themes that were distant from the literal content of the lesson but more familiar to students. Through these discursive moves, meanings were mobilized to interpret the text through references to other topics, although not all of these links were accepted by the students, as evidenced by their silence in relation to the statement that now everyone could own land. The range of associated meanings was set forth but also delimited by the teacher's discourse.

As a model of what it means to read and to understand a written text, this form of mediation implicitly fostered reference to other structures of meaning derived from both previous classroom experiences and everyday life. Through this sort of contextualization, the teacher mediated the practice of reading in a particular manner: meaning can be drawn from other contexts and taken to the text.

Towards a Literal Meaning: What Was France Like?

The second case took place in 1983, with the sixth-grade class in a small rural village, where most parents worked their fields and herded sheep, though many were also weavers. The young teacher of this class, Ramón, was particularly intent upon teaching reading, and would urge his students to read anything they could find. He often modeled by reading the textbook lesson in a slow and even voice. He told me that he always attempted to "draw out" (*extraer*) students' understanding of what they had read, rather than "transmitting" contents to them through verbal lecturing. In this case he was intent on having the children recall the literal content of the lesson. The lesson is on the French Revolution, and begins with a description of France, as follows:

> THE FRENCH REVOLUTION: A REVOLUTION IS IN PREPARATION
> *France is one of the largest countries of Europe and also one of the most populous. It has a temperate climate and it rains abundantly in winter. There are many rivers and the soil is fertile; for this reason, agriculture is an important activity; they grow cereals, potatoes, fruit, grapevine, vegetables. Animal husbandry is also developed, and, keep livestock. Cows give good milk for the production of cheese.*
>
> *In the 18th century, France was a wealthy and powerful country, with a large army. Its king, Louis XIV, had several beautiful palaces built, where splendid festivities were celebrated. Paris, the capital, was the most famous city of Europe. Apparently everything was going well, but only for the aristocrats, the high clergy and some of the wealthy townspeople who lived in luxury. Most inhabitants of the country lived miserably. The peasants, artisans and burghers were the only ones to pay taxes, and in spite of that, they had no rights and no laws to protect them [. . .] Peasants were the poorest social class, they worked hard and, on certain days, served their lords with no pay.*[8]

After students had read the lesson in silence for about 10 minutes, the teacher asked them to close their books and began to ask questions.

1. T: let's see, you must know it [the lesson] by heart now, tell me what you know.
2. Juliana: (begins to answer) The peasants (inaudible) ...

3. T: (interrupts): no, first, what is the lesson about(?) (the students do not answer)
4. T: (provides the answer) the French revolution, it says they are preparing the revolution ... why does it say that(?) (further questioning does not produce answers)
5. T: let's see, Angeles, in France, what was there(?)
6. Angeles: lots of fertile fields, rivers, and sheep (taking livestock in the text to mean sheep)
7. T: and what does it say about the sheep(?) (Angeles does not answer, and other students begin to laugh, perhaps responding to the teacher's intonation)
8. T: (asks Felipe) What do you remember(?)
9. Felipe: Louis XIV wanted France and … (inaudible) (he refers to a later part of the lesson)
10. T: and(?)
11. Felipe: the artisans and burghers …
12. T: (again interrupts) First, let's see what they say France is, is it poor or rich, with much cattle, or little cattle(?)
13. Felipe: lots of cattle
14. Juan: it had many towns
15. T: you have not spoken well, you have not expressed yourselves ... What was the weather like(?)
16. Benito: warm.
17. T: what was produced(?)
18. Ss (several at a time): Fruit, potatoes, grapevine (vid)
19. T: and what is vid? (The students don't answer) … I thought you didn't know … vid are the grapevines, it comes from viñedos (vineyards) where grapes are planted … Have you seen them on television? How they produce wine(?) OK, grapevines … and what else? (students do not answer)
20. T: what was France like then(?) Was it poor, or rich(?)

Students are increasingly reticent even in the face of fairly easy answers, so the teacher changes strategies and announces a written questionnaire, which he proceeds to dictate to the students.

This extract has little variation in its generic texture. The structure of interaction followed the triadic pattern closely: sequences are initiated (I) by a question from the teacher, followed by student responses (or lack thereof) (R), and terminating with implicit or explicit evaluative moves favoring answers taken from the text (E) (in lines 3, 7, 10, 12, 15), at times combined with a new initiation. Contextualization was accomplished through the questions Ramón addressed to the students and the answers that he accepted as valid. After a false start, when he asked them what they knew and Juliana mentioned *campesinos* (peasants), Ramón reframed the question as: "first, what is the lesson about?" He then redirected the children's attention by reading the title and subtitle of the lesson.

The questions he formulated are representative of a speech genre; students recognized the form as known-answer questions, and attempted to give appropriate answers. His interventions oriented students' recall to the sequence of the text and favored literal reproduction of phrases to complete his questions. Interaction did not flow, however, and eventually broke down, as the answers offered did not meet with the teacher's agreement. Students mentioned details that were not central to the first paragraph, such as sheep, Louis XVI, artisans, so Ramón redirected them again, asking what France was like at the time. His attempt to have Angeles say more about the sheep met with laughter from other students, who perhaps recognized that the term was not acceptable, as it was not mentioned in the text. When other students provided answers that were closer to the text (warm, grapevines), the teacher's uptake was still problematic, as when he asked for the meaning of an unfamiliar term, *vid*. This (line 19) is the one turn in which Ramón offered information, rather than requesting it; he glossed the word *vid*, grapevine, as the plant that produces grapes, and then, in the sole intertextual move of this exchange, he referred to vineyards represented in televized wine commercials.

The topic of the lesson, the French Revolution, was distant from the students' experience or previous schooling, as was the topic of the Middle Ages in the previous example. Students attempted to relate the text to themes that were familiar to their own worlds, for example in the first response, "*los campesinos*." When Angeles offered sheep as a synonym for *ganado*, livestock, she may have had in mind the fact that several families in the village owned herds of sheep, but had little experience with cows (*ganado vacuno*).

In this case, unlike the previous one, the teacher's mediation of the text tended to exclude interpretations that were related to the immediate cultural context of the students.[9] Contextualization of the text operated through a verbatim reproduction of phrases taken from the written text. The meaning constructed in this manner located "authority for interpretation of written text in the written text itself" (Bloome and Egan-Robertson, 1993). This approach to texts constructs the illusion of a decontextualization of school knowledge, particularly that conveyed in writing, when in fact the meaning that is constructed is not inherent in the text, but depends equally upon a series of strictly contextual discursive moves, constituting a specific speech genre, which is deployed and recognized by both the teacher and the students.

Enacting the Text: The *Grito de Dolores*

In other classes, the generic texture of the teacher's discourse surrounding textbook lessons is more complex. In the following fourth-grade lesson given during the 2010–2011 school year, Eva, a middle-aged teacher, focused on the "Queretaro Conspiracy," an episode in Mexico's struggle for Independence, and the "*Grito de Dolores*," the call to arms delivered by Miguel Hidalgo (the parish priest in Dolores) on the dawn of September 16, 1810. After asking students to read the lesson, Eva signaled a new genre: "OK, we are going to make a short commentary." While students followed in their textbooks she read short parts aloud, suspending at times to insert brief "commentaries" in which she would describe—and at moments enact—the situation faced by the early insurgents. The following extract illustrates her discourse; the text being read is in italics.[10]

1. T: … OK, we continue (she resumes reading) … *the conspiracy was discovered, but before the authorities could catch the participants, the brave Mrs. Josefa found out and was able to warn Aldama, who rode that night from San Miguel to Dolores to warn Hidalgo and Ignacio Allende … their plans had been betrayed* (ends reading). so … that is before they were imprisoned … what did Josefa do(?) ... she went and told whom(?) (pause) … whom did she tell(?)
2. S: Miguel Hidalgo.
3. T: not true(!) … Aldama … Aldama … what … isn't it in your book(?) (She told) Aldama … and Aldama then had to tell … whom (?)
4. Ss: Miguel Hidalgo!
5. T: Miguel Hidalgo and who else(?)
6. Ss: Ignacio Allende (many students participate simultaneously with different names) …
7. T: Ignacio Allende … ok … he told them (she uses gestures and direct discourse) … 'you know what(?) … they've discovered us(!) … now… now maybe there is no point in continuing(!) so(?) … but what did they do(?) they did not give up(!) ... rather, they …(?)
8. Ss (overlaps): they moved up the date (inaudible … overlap) …
9. T: 'ok … it is time to rise up in …(?)' (voicing the insurgents' call)
10. Ss: arms(!)
11. T: arms … ok (continues to read).

A number of genres are embedded within the basic pattern of triadic discourse, marked by teacher queries, student responses, and teacher evaluations (e.g. turns 3 and 11). However, Eva's discourse at this point had shifted towards a narrative tone, perhaps supported by the style of this particular lesson, as she recounted how the early insurgents' conspiracy was discovered. In turns 7 and 9, Eva in fact enacted the account, by using *direct discourse* and dramatic gestures: "you know what, they've discovered us … maybe there is no point in continuing!" and "it is time to rise up in …?" letting the students add the final word, "arms." In these moves she played the role of Aldama, relaying his voice as he carried the news to the insurgents in Dolores. Eva continued to read the text, and then returned to the dramatic tone asking students to imagine that they were present at the moment of the *Grito de Dolores*:

1. T: (reading) *Now each year we celebrate the* Grito de Dolores, *but on that early morning, the atmosphere was tense* … (suspends reading) logically there was tension, right(?) ... if you had been there … wouldn't you be frightened(?)

2. Ss: (a few) no(!)
3. T: are you sure you would not be frightened(?)
4. Ss: yeees(!)
5. T: I say yes … right(?) "what if they kill us(?) what if we don't gain anything(?)" … right(?) then … first of all … they didn't even have arms … right(?) what were their arms(?)
6. Ss: sticks … slings … stones … machetes … (different voices repeating words)
7. T: sticks … slings … machetes … right(?)

After achieving agreement from the majority (turn 4), Eva again shifted to direct discourse, unannounced in this case: "What if they kill us? What if we don't gain anything?" (turn 5). In a contextualization move similar, yet contrary, to that made by Eufemio, who brought the "feudal lord" to the present-day context of the students, Eva attempted to transport the children back in time to the historical context of the *Grito de Dolores*.[11] Like Ramón, she followed the text closely, yet, unlike him, she recast it as a story in order to stress the dramatic aspects of the historical moment.

Intertextual reference was assured as Eva drew on the strong tradition associated with school lessons on the Mexican Independence, as was true of the first example. She could further rely on the children's familiarity with the events through other sources, as the class followed the bicentenary celebrations of Mexican Independence (1810–2010), which received ample media coverage. Further intertextual connections emerged as Eva recounted the events through the use of direct discourse and gesture; these generic resources are associated with the practice of *cuentacuentos* (storytellers), whose skills have been disseminated through the teacher corps during recent years.[12]

Through this storytelling genre, Eva contextualized Hidalgo's call to arms and stressed the difference between what happened during the original *Grito de Dolores* and the yearly celebration of Independence familiar to the students. Rather than merely representing an event of the past, Eva constructed a context in which the students themselves could "take part." Only after setting the scene of Hidalgo's call to arms, and stressing that the people who responded might be frightened, did Eva weave in the question regarding the arms they carried: sticks, slings, and machetes. She followed up not with a questionnaire, but rather with the suggestion given in the textbook: she asked the children to imagine they were reporters who had witnessed the *Grito de Dolores* and to write up the event for a newspaper of the time.[13]

Conclusions

The present chapter compares the literacy practices and discourse genres in three Mexican elementary school classes, in which history textbook lessons are read. The teacher's mediation of the text in each class is organized in a different way.

In the first case reading is a social, public, oral act, done by one student at a time in front of the group. The teacher anticipates the possible meanings of the text to be read, relating it to other school contents, and engages the whole class in the formulation of answers to comprehension questions. In the second case reading is an individual, silent act, though publicly evaluated by the teacher. The teacher lets students read the lesson alone and then asks them to close their books and recall the content, guiding them through a verbatim recovery of its main points. Mediation is instrumental, centered on formal strategies (sequence, selection of main points), intended to facilitate access to the "literal" meaning of the text. In the third case, reading is led and enacted by the teacher, while students follow in their own textbooks. Mediation is dramatic and recounts the event described in the text by giving voice to historical characters.

The teacher's discourse in each case acquired a distinct generic texture as speech genres and written texts intertwined. The first teacher prefaces the text with an extensive "preamble" (his term) on the possible meaning of the title, "The idea of liberty." In the second case, mediation is achieved through a genre of interrogation that directs the students' attention towards a sequential and literal recovery of the text and offers explanation only in the case of problematic words. The third teacher displays her narrative skills in developing a "commentary" on the text, integrating genres that are associated with a dramatic storytelling tradition, such as direct discourse and the invocation of emotional response.

I highlighted contextualization and intertextuality as dimensions for comparison of the examples. The presence or absence of intertextual references distinguishes the manners in which teachers contextualize the texts. Eufemio uses the "preamble" to relate a lesson on the Middle Ages to Mexican history, drawing analogies between the way serfs became freemen and the traditional theme of "freedom from slavery" associated with the struggle for Independence led by Miguel Hidalgo. Students are asked to imagine a situation where a feudal lord would come and ask them to work on his land in return for part of the produce, suggesting that, in contrast to the present day, they would not be able to own their own land. They are asked to reflect on the injustice of that situation. In the second case, Ramón uses direct questioning to steer student attention towards the phrases in the text that describe France before the French Revolution; an intertextual reference appears when he attempts to convey the meaning of grapevines by reference to televised commercials on wine. In this case, contextualization constructs an illusion of a "decontextualized" text related to a distant place, as the teacher's mediation precludes the association with local realities (peasants, sheep). In the third case, Eva uses direct discourse in retelling the episode of the Mexican Independence described in the textbook. Her commentary conveys the tension surrounding the moment in which the insurgents' "conspiracy" was discovered, so that they were forced to immediately launch the "call to arms" planned for a later date. In this case, contextualization takes the form of asking the children to imagine they were present at the time. Intertextual references in the text, taken up later by the children, relate the original *Grito* to present-day celebrations of Independence Day, where the *Grito* is reenacted in all Mexican towns, with special emphasis during the bicentenary year (2010).

Significantly, students were not totally bound by the teacher's mediation of the text. In each case they participate in the construction of intertextual references and resist the interpretations that are being constructed by the teachers: their participation may take the form of alternative references to the text, but also to knowledge acquired through their previous experiences. It may be a negative answer, even when the generic form of the teacher's question begs a "yes" (as in "wouldn't you be frightened?"). It also appears as silence or reticence to participate in the well-known game of known-answer questions. The teachers respond to student interventions in different ways, ranging from partially accepting divergent answers and reformulating assertions to implicitly discounting non-literal interpretations of the text or explicitly rejecting responses that do not fit into the story.

The relation to the meaning of the text—indeed the meaning of textual knowledge in general—is implicitly taught through the teachers' oral mediation of textbook reading. The first teacher models possible interpretations of texts by recasting them in more familiar contexts; the second treats texts as irrefutable, and meaning as inherent in the text; the third shows that it is possible to transform texts into stories and imagine oneself within them. As in the case of traditional literacy studied by Brian Street, "without the oral tradition in which it is embedded, the writing alone is not 'knowledge': it does not have 'universal' meaning, but only that given it by the context" (Street, 1984, 99). Though meaning is in fact always mediated by oral social interaction in a context, certain practices teach that knowledge is in the text, and can be reached despite the context. Nevertheless, the practice of many teachers reveals a much more complex relationship to the text, in which reference to contextual knowledge contributed by both teachers and students is brought to bear upon the task of interpretation. In the light of these examples, I suggest that the notion that school literacy somehow begets a solitary, silent, decontextualized, and objective relationship to written language is a particular reification of researchers' introspective imagination, not a description of actual social uses of writing in school contexts.

Notes

1 Parts of the first two examples analyzed were included in a previous publication, Rockwell, 1995. Permission to reproduce them was given by the publisher.
2 This does not imply that students do not come up with their own interpretations of the texts they read, a process that is also constant, but not the focus of the present analysis.
3 All names are fictitious.
4 My translation of all texts and extracts, from the original in Spanish.
5 The text being read is from the *Libro de Texto Gratuito, Ciencias Sociales*, sexto grado, México, Secretaría de Educación Pública (1988), 27.

6 Notation conventions are: the colon : announces direct discourse by the speaker named or identified with initials M/T (teacher) or A/S (student); three points ... indicate syntactic breaks and/or short pauses; brackets [....] signal omitted or inaudible discourse; a question mark in parentheses (?) is an inferred interrogative tone; text within parentheses (xxxx) gives non-verbal or contextual information; single quotation marks '' mark direct discourse within a turn. Capitals are used only with conventional proper names. Italics are used to indicate either print or written texts that are being read. Missing periods are intentional.

7 The students' reticence may reflect the fact that not all families owned land in the town, hence the teacher's insistence, "but we can all buy land."

8 The text being read is from the *Libro de Texto Gratuito, Ciencias Sociales*, sexto grado, México, Secretaría de Educación Pública (1988), 32.

9 The possibility of relating the content to their knowledge of the Mexican Revolution was not explored, even though during the same day students had made a drawing of the agrarian leader Emiliano Zapata.

10 The text being read is from the *Libro de Texto Gratuito, Historia*, cuarto grado, México, Secretaría de Educación Pública (2007), 89.

11 These discursive moves have evident chronotopic implications, beyond the scope of this analysis.

12 Storytellers and some teachers were trained to read/tell groups stories from the increasingly available children's literature being published and distributed to schools at the time.

13 This exercise proved to be difficult, as not all students were able to place themselves in the situation, and bracket their knowledge of subsequent events and celebrations of Independence Day: for example, some mentioned fireworks.

References

Bakhtin, M. (1986). *Speech Genres and Other Late Essays*. Austin: University of Texas Press.

Barton, D., and Hamilton, M. (2005). Literacy, reification and the dynamics of social interaction. In D. Barton and K. Tusting (eds.), *Beyond Communities of Practice: Language, Power and Social Context*.Cambridge: Cambridge University Press, 14–35.

Bauman, Richard (1992). Contextualization, tradition and dialogue of genres: Icelandic legends of the Kraftaskáld. In A. Duranti and C. Goodwin (eds.), *Rethinking Context*. Cambridge: Cambridge University Press, 125–45.

Baynham, M., and Masing, H. L. (2000). Mediators and mediation in multilingual literacy events. In M. Martin-Jones and K. Jones (eds.), *Multilingual Literacies*. Philadelphia: John Benjamin, 189–207.

Bloome, D., and Egan-Robertson, A. (1993). The social construction of intertextuality in the classroom reading and writing lessons. *Reading Research Quarterly* 28 (4), 304–34.

Briggs, C., and Bauman, R. (1992). Genre, intertextuality and social power. *Journal of Linguistic Anthropology* 2 (2), 131–72.

Cazden, C. B. (1988/2001). *Classroom Discourse: The Language of Teaching and Learning*. Portsmouth, NH: Heinemann.

Chartier, R. (1993). Du livre au lire [From the book to reading]. In R. Chartier (ed.), *Pratiques de la lecture* [*Reading Practices*]. Paris: Payot.

Edwards, D., and Mercer, N. (1989). Reconstructing context: the conventionalization of classroom knowledge. *Discourse Processes* 12, 91–104.

Lave, J. (1996). The practice of learning. In S. Chaiklin and J. Lave (eds.), *Understanding Practice: Perspectives on Activity and Context*. Cambridge: Cambridge University Press, 3–35.

Lemke, J. (1990). *Talking Science: Language, Learning and Values*. Westport, CT: Ablex.
Nespor, J. (1997). *Tangled Up in School: Politics, Space, Bodies, and Signs in the Educational Process*. Mahwah, NJ: Lawrence Erlbaum.
Rockwell, E. (1995). En torno al texto: tradiciones docentes y prácticas cotidianas [Around the text: teaching traditions and everyday practices]. In E. Rockwell (ed.), *La escuela cotidiana* [*The Everyday School*]. México: Fondo de Cultura Económica, 198–222.
Rockwell, E. (2000). Teaching genres: a Bakhtinian approach. *Anthropology and Education Quarterly* 31 (3), 260–82.
Rockwell, E. (2007). Huellas del pasado en el las culturas escolares [Traces of the past in the school cultures]. *Revista de Antropología Social* [*Social Anthropology Review*] 16, 175–212.
Rockwell, E. (2010). L'appropriation de l'écriture dans deux villages Nahua du centre du Mexique [The appropriation of writing in two Nahua villages of central Mexico]. *Langage et Société* [*Language and Society*] 134, 83–99.
Street, B. (1984). *Literacy in Theory and Practice*. Cambridge: Cambridge University Press.

14

LEARNING ENGLISH IN MEXICO: TRANSNATIONAL LANGUAGE IDEOLOGIES AND PRACTICES

M. Sidury Christiansen and Marcia Farr

The reality of transnational communities in the context of globalization has captured the attention of scholars from a variety of disciplines. The most salient transnational communities in the United States are Mexican, and although sociologists and anthropologists, among others, have documented the experiences of these communities (e.g. Hirsch, 2003; Hondagneu-Sotelo, 1995; Massey, Alarcon, Durand, and González, 1987; Smith, 2006), the language and literacy practices of transnationals figure only incidentally, if at all, in these studies. Farr (2006) and Dick (2006, 2010), in contrast, focus centrally on language practices in transnational Mexican communities, and Farr (1994a, 1994b, and 2000) and Guerra and Farr (2002) explore literacy practices in a transnational community. This chapter more specifically focuses on how transnationalism is shaping parental language ideologies in Mexico that now emphasize the importance of acquiring both oral skills and literacy in English in non-traditional ways. To clarify, we use the term "transnational community" to refer to social networks that live on "both sides of a nation-state border and maintain social, economic, political, and emotional ties that extend" between two countries (Farr, 2006, 5). By extension, we use the term "transnationals" to refer to people who maintain such ties, whether or not they are part of an established transnational community. In some mid-size towns in Mexico, for example, the presence of transnationalism is felt even though the entire town is not linked to a specific site in the U.S. from and to which migrants move, because a sufficient number of people have migrated, and returned, to effect changes in traditional language and literacy ideologies.

Our research site, a town of 15,000 in the sierra between the states of Puebla and Veracruz, fits this profile with a high percentage of returnees from the U.S. Those who migrated to the U.S. but then returned to this town introduced new possibilities that inevitably impacted traditional ideologies, even among those who have never been migrants themselves but who now can imagine themselves living and working in the U.S. In this town the presence of such people intensified a long-perceived link between English and more lucrative employment, changing local parents' traditional ideas and practices regarding 1) the importance of acquiring oral and written English, 2) processes of learning, and 3) bilingualism. Parents now place high priority on the acquisition of both oral skills and literacy in English.

The increasingly transnational atmosphere of this town is transforming the traditional attitudes toward language and literacy held by previous generations, since returnees bring back new ideas and practices from the U.S. These changes not only affect working-class populations but are echoed throughout all socioeconomic and educational levels. Many parents now encourage the acquisition of oral and written English, although at varying levels of intensity, and this affects their expectations of both private and public schools. Ideologies about processes of learning also have changed, especially regarding literacy, moving beyond the traditional emphasis on grammar and the repetition of perfect sentences (often including academic or tourist-like phrases) to a literacy that is pragmatic and utilitarian for the workforce.

Background: Migration and Return to Mexico

Mexicans comprise the largest group of Latino migrants to the United States. Many Mexican migrants to the U.S. are from the states of Jalisco, Michoacán, and Guanajuato (Farr, 2006; Zentella, 2005), comprising an estimated 40 percent of the Mexican population migrating to the U.S. (INEGI, 2005). Many of these migrants do not return permanently to Mexico; instead they help other family members come to the U.S. and continue their lives in their new country. Others maintain strong ties both in their U.S. residences and in their places of origin in Mexico, creating what we term "transnational" communities (Farr, 2006). These transnational Mexicans primarily reside in the U.S., but often return to Mexico for extended periods of time or to celebrate important events such as religious holidays (i.e., *Semana Santa*) or personal ceremonies (i.e., *quinceañera* celebrations). Nevertheless, some migrants *do* return to Mexico to stay, and they are also transnational in the sense that they have lived in a different country for an extended period of time, and they have new intercultural identities which make them act and react to their original environment in different ways (Menard-Warwick, 2008).

Over a period of five years, only 14.6 percent of the Mexican population who left for the U.S. between 1995 and 2000 returned to Mexico (INEGI, 2005). Surprisingly, while most migrants to the U.S. are from the west coast of Mexico, many of the returnees are from east coastal states or southern Mexico. Veracruz is one of the states with a high percentage of returnees (INEGI, 2005). Despite having a high percentage of migrants to the U.S. (approximately 5 percent of Mexican migrants), Veracruz has one of the highest levels (over 12 percent) of returnees (INEGI, 2005). This higher level of returnees creates local changes in a different manner from those enacted by the transnational community previously studied by Farr (2006).

The transnational Mexicans studied by Farr (2006), for instance, are changing economic, political, educational, and cultural practices in both countries. The first generation of these migrants came to the U.S., worked, and then retired to Mexico. Succeeding generations obtained citizenship (if they were not born in the U.S.) and continued as part of the American workforce; the third generation especially increased their educational levels with high school and college degrees. The families used U.S. dollars to build and expand houses in Mexico, buy their own land, and move from subsistence to commercial agriculture (for example, in growing and selling avocados in the state of Michoacán). Transnational Mexicans, then, using U.S. dollars, have improved churches, parks, and public works (such as plazas and libraries) in Mexico, and some have entered politics back in Mexico (i.e. Semple, 2010; Thompson, 2004). Their strong presence in the U.S. and their considerable economic remittances to Mexico have facilitated the creation of institutions to support culture and education and, mainly in Mexico, contribute to the creation of more schools, the acquisition of computers and books, and changes in educational programs (FEDECMI, n.d.).

In Mexican states with lower numbers of migrants and higher percentages of returnees, as in Veracruz, however, returnees create a transnationalism that is enacted on a less visible level. Since there are fewer people from Veracruz living in the U.S. permanently, there are fewer transnational ties between U.S. locations and Veracruz localities, and thus fewer strong large organizations to advocate for families left behind. This, of course, is a changing dynamic. Each year, more people from eastern and southern Mexican states migrate to the U.S., so it may only be a matter of time before they begin making similar changes in their communities. Meanwhile, as the migrant population originating from Veracruz grows, the people who return take with them new knowledge to which others are then exposed. While returnees do re-acculturate to many Mexican customs, the knowledge and experiences they bring back to Mexico have an impact on these traditional customs and beliefs. In particular, the language ideologies of both English language teachers (Sayer, 2007) and English language learners (Whiteside, 2006) change as the realities in more areas are shaped by migration patterns. In this chapter we focus on some of these changing beliefs, in particular, on language ideologies regarding the teaching and learning of oral and written English. In doing so, we present an emerging phenomenon regarding language and literacy in Mexico, in keeping with the overall aim of this volume.

Research Sites

Since our goal is to explore how transnationalism is shaping language and literacy ideologies and practices in Mexican states such as Veracruz, we conducted ethnographic research in two settings within our chosen town of about 15,000 in the sierra between Veracruz and Puebla: an informal language school created by a transnational parent, and a public elementary school piloting the *Programa Nacional de Inglés en Educación Básica* (PNIEB) (the National Program of English in Basic Education). Our focus in these programs is on the language ideologies of parents that underlie their efforts to create learning sites for English or to support their children learning English at school.

"English for Children" is an informal language school created by a transnational parent. The school provides two major programs: a regular school program that meets three hours per day during the school year and a summer program that meets six hours per day in the summer. The school's only teacher is Aracely (all names are pseudonyms), who lives one to two years alternately in the U.S. and Mexico. She has some college education, and, although she has no formal training in education, English as a Second Language (ESL), or language pedagogy, she is Cardio-Pulmonary Resuscitation (CPR)-certified, has babysitting experience in the U.S., holds a substitute teaching license, and is a teacher's aide in the state of California. Aracely has been teaching English in this school for several years. The idea for creating a school evolved from her experience babysitting in Mexico for children who were recent returnees from the U.S. Some of these children's parents asked her to teach the children more English, a goal she herself shared so that her own children did not lose their English while in Mexico.

Aracely first went to the U.S. at the age of 20 and did not know English. Now she is more confident speaking it, having attended community college ESL classes in the U.S. Generally, more than 15 children are registered at her school, ranging in age from 2.5 to 12 years, and no more than 8 children attend at a given time. Even though the program started as a way to maintain the English skills some children had gained while living in the U.S., it slowly shifted to improving the English skills of children who 1) may return to the U.S., 2) may move there later (with a relative), or 3) want to prepare themselves for employment in service jobs in Mexico (i.e. at a hotel or as a tourist guide) or in the U.S. The school has only one room, and the teacher makes most of the materials herself. Books and instructional videos were purchased in the U.S. The children in this program vary in their socio-economic backgrounds from middle class to poor-working class.

In the other research setting, at Allende Elementary School, English is taught as part of the curriculum, rather than as a parent-subsidized extra-curricular subject as at two other local public schools and at the informal language school already discussed. Since 2004, Mexico's Secretary of Education has been piloting the *Programa Nacional de Inglés en Educación Básica* (PNIEB) (the National Program of English in Basic Education). This program consists of the teaching of English in primary schools "to be competent in that language and participate effectively in today's world" (PNIEB, 200X, 15) based on the Common European Framework of Reference (Council of Europe, 2001). Its goal is to prepare students to pass the TOEIC-Bridge exam at the end of primary school. Because English is part of the daily curriculum, the elementary school teacher, Alfredo, is also the English teacher. Although he is not required to teach English (due to the piloting stage of this program), Alfredo believes that the Secretary of Education is providing an excellent opportunity with great materials and wants to take advantage of this. Alfredo has been teaching fifth grade for nine years. He has an elementary education degree as well as a Masters in education, and he is pursuing a Ph.D. in education at a local university. He studied English in high school and college, but he did not learn to speak the language fluently, which he attributes to the fact that his instructors mainly utilized the grammar translation method or the audio-lingual method in which simple sentences were drilled and repeated, as opposed to approaches that emphasize activities to promote active English use and communication. He teaches English at least two hours per week using *Enciclomedia*, a virtual English course for children designed by the Secretary of Education or SEP (the equivalent of the Department of Education in the U.S.). When not focusing directly on English instruction, he frequently incorporates English phrases and sentences into his everyday teaching. The virtual course includes a virtual teacher, virtual materials, and interactive lessons. There are 31 children in his classroom (a typical number in a small town public school), and most come from working- and lower middle-class backgrounds.

Research Methodology

We used ethnographic research methods to gather data for this project. One of the authors (Christiansen) was a participant–observer in both settings for seven weeks during the summer session. She attended class every day in the two settings and sometimes, at the request of the teachers, she taught a lesson. The researcher was part of the class sometimes as a detached observer (she sat in the corner while children were engaged in an activity), sometimes as a participant (completing activities with the students, e.g. a bingo game), sometimes as a grader or juror (she was asked to "correct" children's spelling or pronunciation or she was asked to vote for the best role-play), and sometimes as a teacher. In both settings, the researcher was asked to speak in English at all times so that children had the opportunity to practice English. This created an interesting dynamic between the researcher and the children because she is a native Mexican, but was initially considered a Mexican *de allá* (from over there [the U.S.]) who either preferred not to speak Spanish or could not speak Spanish well. Thus, most of the time children would try to engage in a conversation with the researcher in English, but they switched to Spanish to complete their thoughts or when what they tried to say was too difficult for them to express in English. As part of the data gathering, informal interviews were conducted (and recorded) with the two teachers and five parental units (three mothers and two couples). Field notes were written immediately after each class session or interview, and materials used to teach English were studied. We were not able to obtain copies due to copyright restrictions, but extensive notes were taken.

To analyze the data we searched for evidence of language ideologies, including those related to literacy, in the field notes, recorded interviews, and other collected materials. In what follows we discuss the ideologies evident on the part of the parents.

Parents' Language and Literacy Ideologies

English has an important place in the language ideologies of Mexico. It has been called "the language of the future," as one of the parents noted during an interview. English has long been viewed as the language that would make someone more competitive in the professional job market in Mexico. With increasing migration since the 1960s, however, this ideology has slowly expanded to include opportunities beyond elite jobs and higher education. Parents now perceive an additional need to learn to speak, as well as read and write, English, *por si acaso se quiere ir para allá* (in case they want to go to the U.S.). The option of going to the U.S. is more recent within the list of goals a parent holds for his or her children in this community, yet the constant increase in migration makes it important.

All parents in this study agree on the importance of English, even though they vary in economic background. More affluent parents easily acknowledge the possibility of their children having to encounter English at school in Mexico, since programs for some time have copied U.S. curriculum and even have had U.S. professors and teachers at some colleges and high schools. In contrast, less affluent parents either talk about scholarships abroad or the option that children may want to go and work in the U.S.

Despite these differences, much discussion about the need for their children to learn English focused on employment. The more affluent parents noted that their children may need English literacy skills to keep updated with professional magazines or books in English, working as an agent who is sent to the United States, or as a self-employed person who has to do business abroad. The less affluent parents noted that even in the U.S. if they cannot find employment they can *vender lo que sea* (sell whatever/sell anything) if they can communicate in English with the people who have money. Also, some of these parents believe that, if their children can read and write the basics in English, they can become something else besides *obreros* (workers) and hold a managerial position. These beliefs sharply contrast with the common notion held by many in the U.S. that Mexicans do not want to learn English when they come to the United States. Older generations of migrants may not have anticipated learning English before going to the U.S., but current generations do and, moreover, are proactive in acquiring the language even when they have not yet made a decision to migrate. This new language and literacy ideology is based on knowledge brought by returnees as they recount stories of success and failure sometimes due to language abilities, or to the lack of them.

All parents in this study consider English to be of utmost importance for their children's success. They believe that bilingualism will provide more upward mobility in the workplace or in education, whether it is in their own town, in a larger Mexican city, or abroad. When Josefina's parents were asked how they thought she would use English, they responded:

> *Si no se quiere quedar aquí, Josefina puede trabajar como secretaria para una empresa grande, para los aviones de las que venden los boletos en los aeropuertos, o de esas que reciben a los que llegan acá.*
>
> If she doesn't want to stay here, Josefina could work as a secretary of a major corporation, for the airlines at the ticket counter at airports, or [as a] customs officer who processes the people who come here [Mexico].

Josefina's parents did not complete elementary school, but they nevertheless have a clear idea of how English could open doors for their daughter to leave the small town where they sell *elotes* (corn on the cob) for a living; however, they also have ideas about what she could do if she were to stay in their town. Josefina's father explained:

> *Ella puede hablar con esos que vienen de otros lugares aquí que no hablan español y no entienden lo que les decimos, cómo se llama, de esos que puede llevarlos a ver acá el monte o las ruinas.*
>
> She can talk to those who come here from other places who don't speak Spanish and don't understand what we say to them, what do you call them, those whom she can take to see the mountains or the [archeological] ruins.

Josefina's mother expanded on this:

> *Pues igual ya ve que se van a trabajar de recepcionistas a un lugar en donde llaman de donde quiera pues lo mismo tiene que saber hablar todo eso… pues vienen extranjeros de otro lado y a veces hay muchachos hay que los guías, la guía de turismo, ganan bien. Hay muchos que vienen de varios lados y no hay guías que sepan y no hay quien los puedan ayudar*
>
> Well, you know [young people] go to work as receptionists in a place where they call from everywhere, and for that reason she has to know how to speak all that [languages] … also, foreigners from the other side [the U.S.] come and they sometimes need guides, a tourist guide, they make good money. There are many that come from other places and there aren't guides that know [how to speak English] and there is nobody that can help them.

Josefina's father thus focuses not on the possibilities his daughter has in going abroad, but on the importance of intercultural communication and doing business with people from other countries in her home village. A similar notion is echoed by the mother of Margarita (a 5-year-old attending the informal language school):

> *El inglés es el idioma del futuro, ahora todo ya está en inglés o tiene que ver con los Estados Unidos. Si uno no trabaja allá, trabaja para hacer cosas para allá.*
>
> English is the language of the future; nowadays everything is already in English or has to do with the United States. If you don't work there, you work making things for there.

Margarita's mother refers here to the *maquila* (factory) jobs that prevail in the region. She is acquainted with written labels and instructions, which she sees both at work and also in the environment, and she knows that learning how to read English now has a new, more personal meaning, and it is utilitarian.

Along with this new ideology of language and literacy, there is also a new ideology of language learning. Accompanying the recognition that English is now more palpably needed is an understanding that English is only useful if a person can communicate with it, both spoken and written, either in the context of obtaining better jobs or migrating to the United States. Parents expressed their understanding of English as a language similar to Spanish in the sense that both languages use the same alphabet and some words are the same in both languages (those of Latin and Greek origin), but stated that English is challenging for Spanish speakers in its pronunciation. Thus, many parents undermined the long-term approach of the public schools, which tends to focus more on grammar, translation, and repetition rather than communication activities, a theme raised by Alfredo above. José Carlos's father, whose brother lives in Atlanta, Georgia, explained:

> *Eso de la pronunciación y el vocabulario es importante pues es diferente al español, pues claro una persona que aprende el inglés tiene un tipo de acento pero no importa, sino que pronuncien bien, además el acento no tiene que ver con la capacidad de la persona.*
>
> Pronunciation and vocabulary are important because they are different from Spanish. Of course a person who learns English will have an accent, but it does not matter; what matters is that he has good pronunciation. In addition, the accent has nothing to do with the capacity of the person.

Here this father distinguishes between "accent" that identifies a person's national origin and other aspects of pronunciation that might prevent comprehension. Many parents reported wanting the schools to provide more practice speaking for their children; if they mentioned written English it was to be able to read words and sentences associated with possible employment (e.g. labels, instructions). Interestingly, both the parents and the teachers, in their instruction, emphasized productive (speaking and writing) rather than receptive (listening and reading) language skills, except for parental comments about reading English on the job. One parent described how she had to *meter palancas* (pull strings) to register her child to be with Alfredo, a teacher who is known for making the children speak English phrases in the classroom and who makes them write at least once every week, as part of his communicative approach.

Other parents talked about making a class fun and interesting in order to make children speak and write English, which seems to be an important part of their new ideology of language learning, even if they don't know whether or not their children are making any progress. They believed it is better that they are having fun, rather than being bored, as they were when they were children. If the children are interested, they will talk or write more, as one mother noted. This was the reason many of the parents cited for choosing to enroll their children in the informal language school, despite the English lessons being offered at their elementary school, or the lack of money the families might have. For example, Josefina's parents explained why they decided to enroll her in the language school program:

> *Pues mucha gente viene en una feria en septiembre y vienen extranjeros de por allá a andar acá. Vienen a ver las ruinas por allá por San Andrés y la gente va y les enseña, por lo menos ella habla y entre más habla más aprende y pues es, es lo que le digo a ella. Ella tiene que aprender lo que uno trabaja y llevarlo a cabo, y que no ande inventando, que sepa salir adelante que no se le cierre el mundo. Todas las experiencias sirven todo, es tener ganas de superarse, estudio todo, como le digo, hay veces que no hay dinero así para los fines de semana que trabaje pero si tiene intenciones de hacerlo que no se quede con las ganas de estudiar.*
>
> Well, many people come during town fairs in September and foreigners from over there [U.S.] visit here. They come to see the ruins by San Andrés and people go and they show them, and she [Josefina] at least speaks, and the more she speaks [English] the more she learns and that's what I tell her. She has to learn also what we do, and do it, and not fool around so that she learns how to move upward and succeed. All experiences are good, all you need is to want to succeed, study everything, as I tell her, sometimes when there is no money, so during the weekends I tell her to work, for if she has the intention to study, not to hold back.

Josefina's parents, as already noted, did not finish elementary school and make a living as street vendors selling *elotes* (corn on the cob) at 50 cents each. Nevertheless, her mother believes that by sending her daughter to the language school she will acquire English skills needed for the future, as her earlier comment showed. She also refers to her own experiences with foreigners and encourages her daughter to practice speaking with the tourists who come to the town (previously Josefina's father had suggested that she could even become the guide) to make money to support her own studies. Importantly, this family's understanding of practicing to learn English was shared by most other parents, who even acknowledged that they were learning some English from their children by asking them questions or encouraging them to talk about what they learned at school. In Josefina's case, her parents even bought her a small blackboard on which to write the words she knew and practice the vocabulary at home, so addressing literacy as well as spoken issues.

While these new ideologies of language, literacy, and learning may not be shared by all parents in this town, the unanimity of the parents in this study indicates that transnationalism is stimulating some changes, motivating these parents to enroll their children in the language school or choose the teacher who offers English in the elementary school.

Parents' Ideologies of Bilingualism

Another change is in parents' attitudes towards bilingualism. English contact with Spanish, and its result, was noted by many of the parents interviewed. Many were interested in having their children learn both languages well, yet they expressed negative views toward the linguistic changes in the language of those who had travelled abroad. One father explained that he wanted his children to learn both languages well, alluding to "proper" speaking and literacy in Spanish, i.e. standard orthography in terms of spelling and the use of accents. His use of the term "well" also meant keeping the languages separate and avoiding what he calls *palabritas* (little words), as he explains below:

> *Aquí en el país hay palabritas que hay veces que no sabemos ni qué porque hay veces que la juventud ahora como hablan a veces, y es que a veces no es que haya salido de ellos o de aquí del país porque si salió de aquí pues bueno es nuestra cultura pero hay veces que fueron a trabajar a Estados Unidos y ahí llegan dizque hablando como aquella gente y pues no, no está bien.*
>
> Here in our country there are certain little words that we don't know where they come from, because of the way young people speak nowadays, and sometimes the words didn't even come from them or even the country, because if the word was developed here, well it is our culture, but there are sometimes cases such as those people who go to work in the United States and come back pretending to speak like those people and that is not good.

This comment was complemented by his wife:

> *Es que es diferencia para los extranjeros pero es lo mismo, acá se utiliza el español, pero como dice si se va del otro lado allá tiene que saber ese otro tipo de lenguaje. La mezcla no nos agrada, allá lo de allá y aquí lo de acá, que aprendan bien inglés y que aprendan bien el español.*
>
> Languages are different for the foreigners, but it's the same, here we use Spanish, but if one goes to the other side [the U.S.] one has to know another language. We don't like the mixture [of Spanish and English], what is from

> there should stay there and what is from here should remain here, people should learn English well, and they should also learn Spanish well.

Thus both parents express an aversion to mixing vocabularies from two different languages and cultures, even though this would seem to be inevitable, as is evidenced by many global varieties of English (Kachru et al., 2009). This commitment to keeping the two languages separate is associated with an ideology of monolingualism (Farr, 2011; Farr and Song, in press), according to which "languages" are considered to be autonomous, unified systems with discrete boundaries, and which does not sanction particular varieties of a language characterized by, for example, borrowed words from other language resources. This view may be linked to the fact that the parents in this study are all monolingual speakers of Spanish who view English as a tool and not as a way of living, contrary to what the children's practices and teachers, such as Alfredo, with his communicative perspective, suggest. Moreover, a long history of protecting Mexican national identity from encroaching U.S. influences, whether in language or politics, might reinforce attitudes about keeping Mexican Spanish separate and unmixed.

In spite of these beliefs, however, most parents practice English with their children (in the midst of Spanish communication) and want them to grow up bilingual. For instance, Lety, a mother of one of the children at the elementary school, sometimes tries to brush up her English skills by helping her daughter do her homework (which is mostly writing English sentences or vocabulary words); moreover, to help her daughter practice English literacy, she only allows her to watch movies and cartoons in English, activating the English subtitles to help her learn how to spell new words. Whether or not television aids language development, the parent belief is that it undoubtedly provides more space for English, both spoken and written, in her daughter's life and her own. It is important to acknowledge, however, that not all of the parents in this study were willing to try to help their children with learning even if they knew little English. These parents reported that it was the teacher's responsibility to make children learn and that they would prefer not to do something they couldn't do well.

Parents' Ideologies of Language Learning

This variability among the parents relates to another theme of the interviews. Many parents were disappointed by the paucity of information provided by the school system and the teachers. They often asked during interviews whether or not it was acceptable for their children to watch television in English, how old children should be before sending them to school to learn English (so as not to confuse their Spanish), and if too much exposure to English would have negative effects on their children's development in Spanish. The parents of toddlers especially expressed concerns about development in Spanish when they compared their children with monolingual children of the same age who speak Spanish very well (or English for those recently returned from the U.S.). That is, they perceived less progress in either language in their own (developing) bilingual children, contrary to what research has shown—that monolinguals and bilinguals develop similarly over time (Kendall and Fogle, 2006). This then is probably more due to parental fears than to actual delay in language development, suggesting an ideology of monolingualism. In spite of this, however, five parents reported asking other family members to speak to their children in English (either through the phone, via Skype, or by visiting them). These relatives can be living in Mexico or in the United States, but many of them are reluctant, claiming that their skills are not good and that they don't want the children to learn "bad English" from them. Thus parents shared a strong sense of the importance of having a teacher who truly knows English, especially *si viene de allá* (if the teacher comes from the U.S., either from being from the U.S. or from having lived there).

Another issue is practice. The language school teacher complains that parents do not usually understand what is involved in learning a language. She explains that one must practice in order to acquire language skills, but that only parents who have more time and education seem to do this. For instance, Josefina's parents have a strong sense that they want to provide as much as they can to their children, as they did not have such opportunity themselves when they were growing up. They thus encourage studying, but they often do so at their street stand, which means it is interrupted by the responsibilities of helping parents earn money. A parent of a child at the public elementary school commented that she helps her child with math and other subjects, but if she helps her child with English it is usually with writing because she does not speak it at all. Thus, the idea of practicing is acknowledged but not always enacted ideally.

Conclusion

Parents' language and literacy ideologies about the importance of English, of learning English, of spoken and written language, and of bilingualism have been affected by the increasingly transnational atmosphere of this town, that is, by the ideas that returnees from the U.S. bring back to their relatives and friends who stay in Mexico. The new place that English now occupies has affected members of all socio-economic and educational levels, from street vendors to college graduates, from laborers to entrepreneurs. Additionally, this new ideology regarding the significance of English has changed views of language learning. Parents now place great importance on the acquisition of both oral skills and literacy in English, although at varying levels of intensity. They want schools to emphasize speaking English because doing so will bring their children opportunities at home and abroad; however, they are also very interested in literacy skills, although not in the traditional sense, since they primarily hope their children will learn literacy skills that can be used in employment. Thus, these parents desire for their children a literacy that does not simply conform to grammar, spelling, or other prescribed rules for a language, but a literacy that is pragmatic and utilitarian for the workforce. Finally, their attitudes towards bilingualism have changed. Although they now value having two languages as assets, they still retain aspects of an ideology of monolingualism. That is, they believe the two languages should be kept separate and not mixed, even though the children in both research sites (and the adults when they practice English at home with their children) clearly are not adhering to this belief in practice. Most importantly, however, parents express much desire to learn English, or to have their children learn it, puncturing the widely held belief in the U.S. that Mexicans do not want to learn it. This study, in fact, shows the eagerness with which many Mexicans now approach the learning of spoken English and English literacy.

References

Council of Europe (2001). *Common European Framework of Reference for Languages: Learning, Teaching, Assessment.* Cambridge: Cambridge University Press.

Dick, H. P. (2006). *El norte no se hizo para todos/The United States wasn't made for everyone: imagined lives, social difference, and discourse in migration.* Unpublished doctoral dissertation. Retrieved from ProQuest (AA13225448).

Dick, H. P. (2010). Imagined lives and modernist chronotopes in Mexican nonmigrant discourse. *American Ethnologist* 37 (2), 275–90.

Farr, M. (1994a). En los dos idiomas: literacy practices among Chicano Mexicanos. In B. Moss (ed.), *Literacy Across Communities.* Cresskill, NJ: Hampton Press, 9–48.

Farr, M. (1994b). Biliteracy in the home: practices among Mexicano families in Chicago. In D. Spener (ed.), *Adult Biliteracy in the United States.* McHenry, IL and Washington, DC: Delta Systems and Center for Applied Linguistics, 89–110.

Farr, M. (2000). ¡A mí no me manda nadie!: individualism and identity in Mexican ranchero speech. *Pragmatics* 10 (1), 61–85.

Farr, M. (2006). *Rancheros in Chicagoacán: Language and Identity in a Transnational Community.* Austin: University of Texas Press.

Farr, M. (2011). Urban plurilingualism: language practices, policies, and ideologies in Chicago. *Journal of Pragmatics* 43 (5), 1161–72.

Farr, M., and Song, J. (2011). *Language Ideologies and Policies: Multilingualism and Education— Language and Linguistics Compass*, 5 (9), 660–665.

FEDECMI (n.d). Casa Michoacán FEDECMI [electronic version]. Retrieved January 20, 2011 from http://www.fedecmiusa.com/vide.html.

Guerra, J., and Farr, M. (2002). Writing on the margins: the spiritual and autobiographical discourse of two Mexicanas in Chicago. In G. Hull and K. Schultz (eds.), *School's Out: Bridging Out-of-School Literacies with Classroom Practice.* New York: Teachers College Press, 96–123.

Hirsch, J. S. (2003). *A Courtship after Marriage: Sexuality and Love in Mexican Transnational Families.* Los Angeles: University of California Press.

Hondagneu-Sotelo, P. (1995). Beyond "the longer they stay" (and say they will stay): women and Mexican immigrant settlement. *Qualitative Sociology* 18 (1), 21–43.

INEGI (2005). Mujeres y hombres de México [Women and men of Mexico]. *Biblioteca Digital INEGI: Estadística /población y vivienda / Estudios de género* [*INEGI Digital Library: Statistics/Population and Housing/Gender Studies*]. Retrieved January 20, 2011 from http://www.inegi.org.mx.

Kachru, B., Kachru, Y., and Nelson, C. (eds.) (2009). *The Handbook of World Englishes*. Oxford: Wiley-Blackwell.

Kendall, K., and Fogle, L. (2006, April). Raising bilingual children: common parental concerns and current research. *Research Center for Applied Linguistics*. Retrieved from http://www.cal.org. (Retrieved January 2, 2011).

Massey, D. S., Alarcon, R., Durand, J., and González, H. (1987). *Return to Aztlan: The Social Process of International Migration from Western Mexico*. Berkeley: University of California Press.

Menard-Warwick, J. (2008). The cultural and intercultural identities of transnational English teachers: two case studies from the Americas. *TESOL Quarterly: A Journal for Teachers of English to Speakers of Other Languages and of Standard English as a Second Dialect* 42 (4), 617–40.

PNIEB (2009). Programa de estudio. primer ciclo. Etapa Piloto [Study Program, first cycle. Pilot Stage. Secrataría de Educación Pública. Available from http://basica.sep.gob.mex/pnieb/start.php?act=DocAcademico

Sayer, P. (2007). *Legitimate speakers? An ethnographic study of Oaxacan EFL teachers' ideologies about English*. Unpublished doctoral dissertation. Retrieved from ProQuest (3288008).

Semple, K. (2010, June 1). Immigrant in run for mayor, back home in Mexico. *The New York Times*. Retrieved from http://www.nytimes.com. (Retrieved February 28, 2011).

Smith, R. C. (2006). *Mexican New York: Transnational Lives of New Immigrants*. Los Angeles: University of California Press.

Thompson, G. (2004, July 5). Jerez Journal: Mexico's "Tomato King" seeks a new title. *The New York Times*. Retrieved from http://www.nytimes.com. (Retrieved January 10, 2011).

Whiteside, A. (2006). Research on transnational Yucatec Maya speakers negotiating multilingual California. *Journal of Applied Linguistics* 3 (1), 103–12.

Zentella, A. C. (2005). *Building on Strength: Language and Literacy in Latino Families and Communities*. New York: Teachers College Press.

AFTERWORD

THE THREAT OF A GOOD EXAMPLE: HOW ETHNOGRAPHIC CASE STUDIES CHALLENGE DOMINANT DISCOURSES

David Barton

I have greatly enjoyed reading through this treasure trove of different studies of literacy. They cover many areas of learning and life and raise so many different issues. What binds them together so strongly is the framing of Literacy Studies and the particular lens it provides. The studies have in common the basic concepts of literacy studies such as notions of literacy events and literacy practices. They also draw attention to the issues which are commonly highlighted in studies seeing literacy as a social practice, such as the importance of networks of other people in literacy learning, that there are specific vernacular ways of learning, and that the dynamics of literacy practices vary in different domains of life—there are distinct literacies in different domains and specific literacies for learning. These studies are all revealing examples of situated literacies.

Here I want to underline the importance of each of these studies as a distinct situated case of literacy, but at the same time I am struck by how they reach out and make global links and can be part of global discussions which can help people understand and have some control over their lives in a rapidly changing and uncertain world.

The Threat of a Good Example

The threat of a good example is a phrase which I came across in the early 1980s. It was used at that time to describe Nicaragua when discussing social developments there including the national literacy campaign. In fact Oxfam published a book in 1985 entitled *La amenaza del buen ejemplo* about its social projects supporting the poor after the devastating 1972 earthquake and the later civil war. The phrase has been used in many other situations since then. The threat is that if a small poor country like Nicaragua can improve life for the poorest, then why can't bigger, richer countries? A well-thought-out individual case can have a striking effect just by being a good example. This applies to many areas of life, not just to social development.

Applying this to literacy studies, we can see instances where a particular example can be used more broadly to challenge dominant and naturalized discourses. Situated studies of literacy practices can challenge dominant skills discourses of learning and assessment, for example. Each of the studies in this book is a detailed example, something very specific and located in the particularities of its social context. I am interested in the power of these individual cases.

The second chapter of the book, Marildes Marinho's detailed study of the admission of "indigenous people" into the university, is one such good example. It is set in a discussion of problems with some basic sociolinguistic terms, including the conflicting possibilities for the word "literacy" in Portuguese, as well as the words used to describe the indigenous people themselves. The chapter utilizes a literacy studies lens to identify the challenges to change which universities, as the powerful dominant institutions, face when addressing the needs of poor disenfranchised indigenous groups.

The chapter by Elisa Cragnolino examines the plurality of practices in written culture and how they impact on rural families in Argentina. It documents the different literacies people encounter and the different needs for literacy as they migrate to cities in search of work. It is notable how the different studies draw on different aspects of social theory and socio-cultural approaches to language and learning to complement the lens of literacy studies. Marinho brings in French literacy theorists and Cragnolino uses Bourdieu's work for broader framing. The next chapter, by Enna Carvajal, uses multimodality to analyze high school students' digital texts. Similarly, in Chapter 6 Maria del Carmen Lorenzatti reports a detailed study of an illiterate woman in Argentina and how she draws on her multimodal resources to "read reality." Already in a few chapters we have covered different levels of analysis, examining particular communities, as well as focusing on families and looking at individual students. This chance juxtaposition of different studies is rewarding and I pause to wonder about the multiple meanings of "literacy" in these different contexts.

The chapter by Judy Kalman provides yet another distinct context, a workplace where fishermen at sea are grappling with new technologies which can help them in their livelihoods. The study pays detailed attention to what the fishermen actually do out on a boat searching for fish and how they were willing to incorporate the technology of GPS into their activities when they could see how it could serve their purposes. This work is situated within socio-cultural theories of learning. Various themes run through the book. Another chapter also examines issues around how people incorporate new technologies into their social practice. This is the chapter by Irán Guerrero, which examines a very different space: how teachers make use of new technology in their classrooms.

Numeracy is framed in a similar way as literacy—as a social practice—with attention to new mathematical tools needed when people begin trading outside their existing community, in the chapter by Irma Rosa Fuenlabrada and María Fernanda del Prato. Chapter 10, by Gelsa Knijnik, also focuses on numeracy. It explores mathematics education within the landless movement of Brazil and its relation to school maths. Intriguingly, one chapter documents a project which failed and outlines some of the reasons for failure. This study of the development of a community garden, by Shirley Brice Heath and Daniel Sobol, is certainly "a good example," and much can be learned from the failure. The lens of the ethnographer's eye assumes that everything makes sense from the participants' perspective, and failure can be as instructive as success. Related to this, the last chapter, by M. Sidury Christiansen and Marcia Farr, investigates from the participants' perspective whether parents and children in the particular context of Mexican migrants back in Mexico actually want to learn English.

Other studies juxtapose different theories of literacy in different domains of life, so Maria Lucía Castanheira provides clear case studies of children in working-class families "playing school" at home. Similarly, Gloria Hernández investigates young rural students who have only schooled notions of literacy to draw upon when making sense of their everyday lives. And Elsie Rockwell examines the oral mediation of written texts within elementary school classrooms in Mexico.

Global Impact of Local Examples

These chapters work well as specific situated examples of literacy as a social practice. But at the same time they resonate with other specific studies across the world. I will provide just one example, although there are many possible ones to pursue in the book. The study by Kalman of the fishermen learning to use GPS technology set me thinking in many different directions. One of my favorite academic books is John Singleton's 1998 collection *Literacy in Likely Places*. This is a set of studies of different forms of apprenticeship in Japan, ranging from learning to become a potter or a doctor, through to becoming Zen master or a diver. The book is about the commonalities of learning in these "likely places." Keeping to the nautical theme of Kalman's study, I am also reminded of Hutchins' studies of learning to navigate (1995) with its close attention to the practical details of people's decision making in practice. Kalman's work brings these examples of learning in practice up to date with the incorporation of new technologies into everyday practices. And a study based in East Africa (Pfaff, 2010) of people taking up the affordances of mobile phones has many parallels and provides another global link.

Whether or not they are explicit about it, many of the studies reported in this book can be seen as drawing upon ethnographic approaches for their methodology (whilst not claiming to be full ethnographies—see Barton, 2011). They have the attention to close detail of ethnographic approaches, the studies are located in their cultural context, and often participants' perceptions are central. Several are innovative and imaginative in their multi-method approaches. These are all key characteristics of studies utilizing ethnographic approaches. A central question is how one can move out from individual ethnographic studies—or any qualitative research for that matter—to engage with a broader literature and to say something of more general significance. I am sure that the particular cases of each of the chapters can be taken out from their situated contexts in this way and located alongside cases from around the world. In this way the studies are both local and global.

Is there a Latin American Literacy Studies?

The question that arises for me having read these cases is whether there is a distinct Latin American approach above and beyond the particular cases. To answer this I have been thinking about what I knew about Latin American approaches to literacy before reading this collection and how these few threads are woven into something stronger by all these new examples. There have been Specific earlier studies set in Latin America which have been informed by literacy studies, such as Wogan's study of the place of writing in an indigenous community in Ecuador (2004) and the study of scribes and their clients in Mexico City by Kalman (1999). Within literacy studies, the most active area where the voice of Latin Americans has been heard, however, is through studies of Spanish-speaking children and adults in the United States, often making links between home practices and educational practices; for example some of the studies reported in collections such as Hull and Schultz (2001), Perez (2004), Farr (2005), Lewis, Enciso, and Moje (2007), and Purcell-Gates (2007). A study of "illegal aliens" in the United States and how they used their knowledge of Spanish to invent a writing system for English, their "illegal alphabet" (Kalmar, 2001), is a study which deserves wider recognition within literacy studies, showing the centrality of social practices even when discussing phonological aspects of language.

I personally knew about Latin American literacy work first through knowing about the Nicaraguan Literacy Campaigns of the 1980s carried out in Spanish, local languages, and English. At that time there was also information available about the popular education projects in Mexico. I also knew about the theory and the practice of Freirian approaches to literacy emanating from Brazil. Often they were seen as voices from afar, and I remember listening to Paulo Freire speaking in a cold converted church in Edinburgh and advising Scottish adult education projects. Close attention was being paid to how to learn lessons from projects carried out in Brazil. Coming rapidly to the present, in my current research I have been in contact with Latin Americans in Argentina and Mexico as informants in studies I have been doing of writing on the Internet: these photographers strategically make language choices when writing on the Internet, choosing Spanish or English to assert their identities, often wanting to both strengthen local ties and to assert global identities (Lee and Barton, 2011).

These are a whole range of quite disparate studies. They provide individual threads and, from my position at least, ones that are quite hard to bring together into a coherent pattern. This book provides much more ranging over the domains of family, community, education, and work places. The studies of education go from early years through to academic writing in universities. With all these topics held together with a common framework, the strength of Latin American Literacy Studies (or as the editors will rightly insist Latin American Literacy and Numeracy Studies) is assured. There is a parallel with the interaction between Francophone and Anglophone literacy research, and I have recently been involved in a collaboration with French researchers, linking across cultures and traditions of research (Barton and Papen, 2010), and there the studies were both distinctive and situated and at the same time global.

So I want to answer the question "Is there a Latin American Literacy Studies?" with a resounding "Yes." At the same time I want to argue almost the opposite, that, as well as being specific, such studies are inevitably global. To understand how specific studies are both situated and global, I was struck by a recent study of Mexican lives which challenges the myth of "illiteracy" by positioning people in the era of global capitalism (Hernandez-Zamora, 2010). It demonstrates how studies can be very specific but at the same time can locate people in global processes. Locating our studies in these broader processes of globalization can provide clarity for literacy studies. Starting from the idea of "really useful knowledge," Elsa Auerbach (2005) puts forward some principles of how literacy studies can "think globally and act locally." This applies to the studies in this book. These situated studies contain commonalities with studies from around the world and quite similar things are being acted out in very different places. Every chapter is relevant to global concerns. Whether we are thinking of the impact of new technologies, or divisions between everyday knowledge and school knowledge, changing demands with migration to cities, or access to education for excluded groups, these are all literacy issues and they are all global issues. These detailed studies contribute to profound discussions about the nature of knowledge and communication in a changing world which looks increasingly risky, vulnerable, and uncertain.

References

Auerbach, E. (2005). Connecting the local and the global: a pedagogy of not-literacy. In J. Anderson, M. Kendrick, T. Rogers, and S. Smythe (eds.), *Portraits of Literacy across Families, Communities, and Schools: Intersections and Tensions*. Mahwah, NJ: Lawrence Erlbaum, 363–79.

Barton, D. (2011). Ethnographic approaches to literacy research. In *The Encyclopedia of Applied Linguistics*. Hoboken, NJ: Wiley-Blackwell.

Barton, D., and Papen, U. (eds.) (2010). *The Anthropology of Writing: Understanding Textually Mediated Worlds*. London: Continuum.

Farr, M. (ed.) (2005). *Latino Language and Literacy in Ethnolinguistic Chicago*. Mahwah, NJ: Lawrence Erlbaum.

Farr, M. (2006). *Rancheros in Chicagoacán: Language and Identity in a Transnational Community*. Austin, TX: University of Texas Press.

Hernandez-Zamora, G. (2010). *Decolonising Literacy: Mexican Lives in the Era of Global Capitalism*. Bristol: Multilingual Matters.

Hull, G., and Schultz, K. (eds.) (2001). *School's Out! Bridging Out-of-School Literacies with Classroom Practice*. New York: Teachers College Press.

Hutchins, E. (1995). *Cognition in the Wild*. Cambridge, MA: MIT Press.

Kalman, J. (1999). *Writing on the Plaza: Mediated Literacy Practices among Scribes and Clients in Mexico City*. Cresskill, NJ: Hampton Press.

Kalmar, T. (2001). *Illegal Alphabets and Adult Biliteracy*. New York: Routledge.

Lee, C., and Barton, D. (2011). Constructing global identities through multilingual writing practices on Flickr.com. *International Multilingualism Research Journal* 5 (1), 39–59.

Lewis, C., Enciso, P., and Moje, E. B. (eds.) (2007). *Reframing Sociocultural Research on Literacy: Identity, Agency, and Power*. Mahwah, NJ: Lawrence Erlbaum.

Pérez, B. (ed.) (2004). *Sociocultural Contexts of Language and Literacy*, 2nd ed. Mahwah, NJ: Lawrence Erlbaum Associates.

Pfaff, J. (2010). A mobile phone: mobility, materiality and everyday Swahili trading practices. *Cultural Geographies* 17, 341–57.

Purcell-Gates, V. (ed.) (2007). *Cultural Practices of Literacy: Case Studies of Language, Literacy, Social Practice, and Power*. Mahwah, NJ: Lawrence Erlbaum.

Wogan, P. (2004). *Magical Writing in Salasaca: Literacy and Power in Highland Ecuador*. Boulder, CO: Westview Press.

ABOUT THE EDITORS AND CONTRIBUTORS

David Barton, Ph.D.
Professor
Lancaster University Department of Linguistics
Lancaster, UK

Enna Carvajal, Ph.D.
Director of Teaching with Technology
Laboratorio de Innovación en Tecnología Educativa
México D.F., México

Maria Lucía Castanheira, Ph.D.
Profesora de la Facultad de Educación de la UFMG
Universidade Federal de Minas Gerais. CEALE/ Faculdade de Educação
Belo Horizonte, Brazil

M. Sidury Christiansen, MA
Ph.D. Candidate
School of Teaching and Learning
Ohio State University
Columbus, Ohio, USA

Elisa Cragnolino, Ph.D.
Associate Professor
Centro de Investigaciones de la Facultad de Filosofía y Humanidades
Universidad Nacional de Córdoba
Argentina

María Fernanda Delprato, MS
Assistant Professor
Centro de Investigaciones de la Facultad de Filosofía y Humanidades
Universidad Nacional de Córdoba
Argentina

Marcia Farr, Ph.D.
Professor
School of Teaching and Learning
Ohio State University
Columbus, Ohio, USA

Irma Rosa Fuenlabrada, MS
Associate Professor
Departamento de Investigaciones Educativas, Centro de Investigación y Estudios Avanzados del IPN
México D.F., México

Irán Guerrero, Ph.D.
Associate in the Laboratorio de Educación, Tecnología y Sociedad
Centro de Investigación y Estudios Avanzados del IPN
México D.F., México

Shirley Brice Heath, Ph.D.
Professor of Linguistics, Emeritus
Stanford University
Stanford, California, USA

Gloria Hernández, Ph.D.
Professor
División Académica Ecatepec
Instituto Superior de Ciencias de la Educación del Estado de México
Estado de México, México

Judy Kalman, Ph.D.
Professor
Departamento de Investigaciones Educativas,
Centro de Investigación y Estudios Avanzados del IPN
México D.F., México

Gelsa Knijnik, Ph.D.
Professor
Universidade do Vale do Rio dos Sinos
Porto Alegre, Brazil

María del Carmen Lorenzatti, Ph.D.
Associate Professor
Centro de Investigaciones de la Facultad de Filosofía y Humanidades
Universidad Nacional de Córdoba
Argentina

Marildes Marinho†, Ph.D.
Profesor at the Facultad de Educación de la UFMG
Universidade Federal de Minas Gerais.
CEALE/ Faculdade de Educação
Belo Horizonte, Brazil

Elsie Rockwell, Ph.D.
Professor
Departamento de Investigaciones Educativas,
Centro de Investigación y Estudios Avanzados del IPN
México D.F., México

Daniel Sobol, BA
Artist
Continuum Corporation
Boston, USA

Brian Street, Ph.D.
Professor
Kings College
London, UK

INDEX

Page references in bold refer to a table and those in italics refer to a figure.